Mark Grimshaw

The Acoustic Ecology of the First-Person Shooter

Mark Grimshaw

The Acoustic Ecology of the First-Person Shooter

The Player Experience of Sound in the First-Person Shooter Computer Game

VDM Verlag Dr. Müller

Imprint

Bibliographic information by the German National Library: The German National Library lists this publication at the German National Bibliography; detailed bibliographic information is available on the Internet at
http://dnb.d-nb.de.

Cover image: www.purestockx.com

Published 2008 Saarbrücken

Publisher:
VDM Verlag Dr. Müller Aktiengesellschaft & Co. KG , Dudweiler Landstr. 125 a,
66123 Saarbrücken, Germany,
Phone +49 681 9100-698, Fax +49 681 9100-988,
Email: info@vdm-verlag.de

Produced in Germany by:
Reha GmbH, Dudweilerstrasse 72, D-66111 Saarbrücken
Schaltungsdienst Lange o.H.G., Zehrensdorfer Str. 11, 12277 Berlin, Germany
Books on Demand GmbH, Gutenbergring 53, 22848 Norderstedt, Germany

Impressum

Bibliografische Information der Deutschen Nationalbibliothek: Die Deutsche Nationalbibliothek verzeichnet diese Publikation in der Deutschen Nationalbibliografie; detaillierte bibliografische Daten sind im Internet über http://dnb.d-nb.de abrufbar.

Coverbild: www.purestockx.com

Erscheinungsjahr: 2008
Erscheinungsort: Saarbrücken

Verlag: VDM Verlag Dr. Müller Aktiengesellschaft & Co. KG , Dudweiler Landstr. 125 a,
D- 66123 Saarbrücken,
Telefon +49 681 9100-698, Telefax +49 681 9100-988,
Email: info@vdm-verlag.de

Herstellung in Deutschland:
Schaltungsdienst Lange o.H.G., Zehrensdorfer Str. 11, D-12277 Berlin
Books on Demand GmbH, Gutenbergring 53, D-22848 Norderstedt
Reha GmbH, Dudweilerstrasse 72, D-66111 Saarbrücken

ISBN: 978-3-639-02408-1

Acknowledgements

Whilst this work was researched and written over a period of three years, its gestation began several decades previously and may be dated back to 1978 when my parents were wise or foolish enough (depending on your point of view) to buy me a secondhand Atari VCS console. For the next few months, the acoustic ecology of my life after school was augmented by the blip.......blip.....blip...blip.blip of Pong. Soon, I was spending to the limits of my pocket money in the dark, seedy dens of gaming arcades or in hotel lobbies in front of the altar of Space Invaders where my shoulders and head were bent for many hours in reverential concentration. To my parents, then, this work is dedicated: something other than a 'complete waste of time and money' (again, depending on your point of view) may arise out of an apparently misspent youth.

A debt of gratitude is also owed to those pale coders who invented the First-Person Shooter and re-introduced me to the fun and passion of computer gaming after the boredom, frustration and disc-swapping of RPGs on the Atari ST. The latter may have nearly killed my will to play but the former rekindled it to such an extent that it continues to burn bright to this day.

My heartfelt thanks to Mireille Ribière who, while on holiday, and at short notice, inserted so many editorial glyphs that the returned work was more a palimpsest than anything the original author might recognize. Syntactical and grammatical errors remaining in this version are due to my negligence alone.

Finally, as this work is based on my PhD thesis, I must thank my supervisors: Sean Cubitt, Craig Hight, Gareth Schott and Bevin Yeatman. Being presented with often wildly divergent suggestions, opinions, methods and ideas may have been, at one and the same time, frustrating, bewildering, amusing and invigorating but I hope I have been able eventually to chart a successful course through those rocky reefs. Any shipwrecks to be found in this work are solely my fault.

Table of Contents

Illustrations and Tables

Illustrations

Tables

Chapter 1

Introduction

1.1 Prelude

The vista that opens up in front of you is of an abbey complex comprising peaceful arched cloisters, verdant lawns, flowerbeds and trees, marble statues, stone steps and pathways that thread their way between a mix of red-tiled Mediterranean-style buildings and Gothic architecture of imposing proportions all set against the backdrop of a blue summer's day. Birds twitter in the background and, to your right, someone is playing a Bach fugue on a church organ. Other members of your platoon cluster around you and over the radio comes the message *Follow me!* with others responding *Affirmative!* You decide to follow the organ music discovering that it emanates from a Gothic church with buttressed superstructure. In the distance, the sharp crack of gunfire and the dull thud of explosions catch your attention. Eager to join the fray, the metronomic rhythm of your running boots on the hard surface of the path is soon matched by the sound of your panting breath; this quickly overtakes the pace of your slowing footsteps so you decide to take a breather and slow to a walk. Soon the organ music is left behind and the cacophony of battle intensifies. Suddenly, a siren indicates that a platoon member has managed to steal the enemy's flag and, amidst the sharp staccato of machine gun fire, you see and hear him weaving and running towards you, flag in hand, hotly pursued by a posse of enemy soldiers. Quickly switching weapons, you lob a few grenades behind him, hearing their clattering skittering on the hard surface of the path, and immediately switch back to your trusty SPAS shotgun. A booming grenade explosion accounts for some of the enemy but more are upon you and your flag carrier slows to a crawl, frantically radioing *Medic!* as he groans in pain leaving a trail of blood behind him. Moving to block a looming enemy combatant you fire a shotgun round at her but miss. The wait for the shotgun to pump another shell into the chamber seems to take an eternity but soon you are able to get off another round and you note, with a guilty pleasure, her screams and the satisfying crump of her body as it falls to the ground. By this time, the patter of footsteps behind you combined with an increasing density of firepower has signalled the arrival of your platoon so you move to medic the flag carrier; while slowly applying bandages, bullets ricochet off the stonework around you, zing off your helmet and, with a grunt, you take one in the leg. Ignoring the life-blood seeping from your own wound, your medical ministrations shortly stop the moaning of your teammate and restore his health and soon he is able to continue to your base where, with a musical flourish, the flag is captured. Kneeling while you reload your shotgun and accept a bandage from a medic for that bullet hole in your leg, you bask in the glow of the radioed congratulations of your platoon. Sir Fragalot,[1] yet again, saves the day![2]

---

1 The moniker I typically play under.

2 Meanwhile, the organist is once more playing that fugue.

1.2 The topic

The above scenario, taken from a First-Person Shooter (FPS) game in capture the flag mode[3] is useful in several respects when laying out the groundwork for this work. It intimates a level of engagement with the game signaling, among other things, player enjoyment and satisfaction at being involved in a team situation that nevertheless rewards individual skill. This pleasure results from (amongst other possible reasons) a demonstration of individual skill[4] combined with the pursuit of collaborative objectives in an environment that, in many respects, simulates real-world problem-solving scenarios. It suggests that such team and individual skills are learnt and acquired on the basis of both real-world and game experience. It shows that point of view and the nature of engagement with the game is subjective; the plot of this story would be very different if narrated by another player. It indicates that the FPS game typically places the player in a hostile environment (the hunter and the hunted) which demands that the player be attentive to all available cues (especially, and crucially for the aims of this work, to sound cues) for team success, character survival and individual glory. And it illustrates the first-person perspective and immersion of the player in the FPS game world. What is described is not fiction or imagination. It is a lived account of experiences within the immersive space of a particular digital game which, in this case, is a form of virtual embodiment strongly associated with the FPS game.[5]

Most of the points raised above may be applied to many digital games[6] with varying degrees of success and prioritization. Some games provide different cues for the solution of a puzzle, some have less of a team aspect or none at all, others are less combative while others offer different perspectives.[7] Broadly speaking (an enlarged definition of the FPS game is provided in section 1.4), it is the types of cues offered, the hostile environment, the mix of team and individual skills,[8] the immersive, first-person perspective and, of course, the combat that signals this type of game as an FPS game. As regards the use that may be made of sound, it is my contention throughout this work that sound cues in the FPS game afford more possibilities than in other digital game genres to live the type of game experience described above. As a first- person perspective game, the FPS game uses sound to immerse the player within the game environment in a way that a 2-dimensional platform game such as *Donkey Kong* (Miyamoto, 1981) or a variety of Role Playing Games (RPGs) do not attempt; these games use other cues (image or text) and sound, I would suggest, is relegated, in the main, to the role of sensory filler.[9]

---

[3] Opposing teams score points by capturing the enemy's flag.

[4] In a multiplayer situation, a public demonstration of such skill.

[5] I have also deliberately used the second person singular not merely in an attempt to engage the reader but because this is the mode of address used by FPS game marketers to the potential purchaser of the game; from the manual for *Doom 3*: "You are a marine [...] Only you stand between Hell and Earth" (id Software, 2004). This is the first stage of immersion in the FPS game world.

[6] Not to mention other forms of gaming.

[7] For a more complex and extensive attempt at taxonomizing digital game types, see Aarseth, Smedstad and Sunnanå (2003).

[8] Although many FPS games offer the deathmatch mode where individuals compete solely against each other while other FPS games pit a lone player against a range of digitally-generated foes.

[9] This is a broad statement that serves to underline the difference in sound use between the FPS genre and other genres. There are, of course, exceptions such as the thoughtful use of sound in *Myst* (Cyan, 1993); the solving of some of that game's puzzles does require that attention be paid to sound and many games, particularly of the platform genre, use sound to inform the player that an item has been picked up or accessed.

This work is about sound in the FPS game and the player relationship to that sound and presents the hypothesis that sound in the FPS game is an acoustic ecology in which the player is an integral and contributing component. The use of the word 'ecology' presupposes the notion that there is a web of interaction occuring (on a sonic level in this case) and that this sonic interaction comprises the acoustic ecology. The hypothesis suggests that, in large part, players make use of sound to navigate through the game world and to interpret events as they occur in the game. In other words, there are responsive relationships between players (in a multiplayer game) and between the players and the game engine that are expressed and mediated through sound.

In computer game studies, it is quite common to come across the term *the game environment* or descriptions of the game world or a virtual world as an *environment* (and such is used terminology throughout this work); examples of such usage may be found in a range of authors (Bridgett, 2003; Larsson, Västfjäll, & Kleiner, 2001; Law, n.d.; McMahan, 2003; Murphy & Pitt, 2001; Stockburger, 2003; Whalen, 2004; Wolf, 2001). A description of an environment, though, while the environment itself may have been affected or even effected by organisms from termites to humans, says little if anything about the relations between that environment and its inhabitants if the dictionary definition of environment is used: "[T]he surroundings or conditions in which a person, animal, or plant lives or operates" (Pearsall, 2002). This then is the sense in which the term environment is used in this work, particularly in relation to a snapshot or series of snapshots of the physical, visual structure of the game level as displayed on the screen. Thus, the environment of the *Urban Terror* (Silicon Ice, 2005) level *Abbey* comprises, among other structures, a blue sky, green grass lawns, a church, cloisters, room interiors, steps and marble statues.

However, environment in this sense, and as exemplified by the level description above, says nothing about the organisms (virtual organisms in this case) which inhabit it nor does it say anything about relationships between those organisms and their environment. To do this, it is necessary to study the game level as an ecology. The scientific definition of ecology decribes a study or an area of knowledge and acoustic ecologists define the term *acoustic ecology* as concerning "the *relationship* between soundscape and listener" (Westerkamp, 2000, p.4). In the case of the acoustic ecology of the FPS game, then, it is the set of relationships between the player(s) and the soundscape(s) of the game.[10] I further suggest, because the sound heard by any individual player in a multiplayer FPS game depends, in part, upon the actions of other players, that the acoustic ecology of a multiplayer FPS game also includes a set of relationships between players founded upon sound. Including the actions of players throughout gameplay encompasses the role of time in helping to create the ecology or to define and refine it — as argued in this work, time is a fundamental component of sound and thus any discussion of an acoustic ecology must take into account time. Viewing the world of the FPS game as an ecology rather than an environment allows for a holistic view of the game of which the acoustic ecology is a part.[11]

---

[10] Ecology is also often used more directly and less conceptually both in popular usage and by ecologists themselves: "We are not outside the ecology [...] we are always and inevitably a part of it" (Bateson quoted in Westerkamp, 2000, p.3). This quote is given, and accepted uncritically, within the same body of writing by Westerkamp from which the previous quote is taken. Within this thesis, I use the term acoustic ecology in the second sense as a synonym for acoustic ecosystem.

[11] Other researchers may be interested in defining and investigating the whole game as a vital ecology — game genres such as Massively Multiplayer Online Role-Playing Games (MMORPGs) may be more fruitful areas of research

This acoustic ecology is not fixed or static. It is constantly changing as players respond to sounds from other players (or computer-generated characters) with their own actions, thereby contributing additional sounds to the acoustic ecology and potentially providing new meaning to, and eliciting further responses from, other players. Furthermore, there are other responsive relationships whereby players respond to sounds produced by the game engine while the game engine itself produces sounds in response to player actions. All sound-producing game objects and events are contributing components; this includes players who, through control of their game character, actively participate in the construction of this acoustic ecology.

In the context of real-world acoustic ecologies, the ecology may be defined "as the relationship between *quality* of an environment and people's state-of-being inside that environment" (Böhme, 2000, p.14).[12] This quotation presents two important concepts that are developed further for the purposes of elucidating the acoustic ecology of the FPS game. It describes what the basic components of an acoustic ecology are — an environment and people — and it suggests that the ecology itself may be described as a relationship between those components. Furthermore, Böhme suggests that this relationship leads to what he calls atmospheres: "Atmospheres stand between subjects and objects, [they are] subjective, insofar as they are nothing without a discerning Subject" (p.15). Atmospheres may be tense, depressing or light, for example; the atmosphere of the typical FPS game may be described as a combination of suspense and threat and, using Böhme's statement above, is the result of the relationships in the game's acoustic ecology.[13]

Ecology as a biological study defines abiotic and biotic components of an ecosystem. For example, a natural ecosystem, such as my garden as I write this, has abiotic components, sunlight and pollution, and biotic components like birds and insects. If this were to be defined in acoustic terms only, I would be hard-pressed to discover its abiotic components because abiotic sounds, particularly in an urban environment such as this one I am in, are rare. If there were the sounds of rustling wind, crashing waves or an exploding volcano then these would be abiotic sounds yet, at this point in time, I hear only biotic sounds; sounds which are evidence of life and, more importantly, lifeform activity. Even the distant sound of driving cars is properly described as a biotic sound because, although mechanical in nature, ultimately it is derived from human and, therefore biological, activity.

Similarly, if sounds produced by in-game characters in the FPS game are described as biotic (in a virtual sense), there are few abiotic sounds to be found. They do exist, such as the sound of dripping water in a cavern in the *Urban Terror* level *Paradise*, but most FPS sounds are biotic. They are derived from and are evidence of in-game biological (player) activity. In the case of character-driven sounds in a multiplayer game configuration, they are evidence of real human activity and, consequently, are evidence of human presence and immersion within the game's acoustic ecology (and, therefore, within the game world). This leads to an important statement about sound usage in the FPS game and one which is formulated and explored in chapters 5 and 7: sound in the FPS game is generally representative of character

---

in this respect than FPS games.

12 This quotation is from the editor's introduction to the article.

13 Not to neglect the contribution to such an atmosphere of other factors such as light and shade in the screen image, for example, or the *prima facie* nondiegetic music.

and, therefore, player activity. Furthermore, it is typically the case that the abiotic components of the game level are sonically reified through parameters of a biotic sound (such as the reverberation cues in footsteps, for example, providing information about the game environment).

Anderson (1996) suggests that our perceptual systems evolved to gather information from natural ecologies and that, in the context of cinema, the senses will interpret the data they receive as if such data were derived from a natural ecology. Similar arguments have been suggested by Breinbjerg (2005, p.3) and Gaver (1993); the former suggesting that the sound of the FPS game should be listened to with our 'natural way of listening' while the latter suggests that the design of auditory interfaces for computer systems should emphasize our 'everyday listening' (see chapter 4). Anderson, though, suggests that rather than adopting such methods of listening when confronted by the unnatural world of the cinema, we (continue to) listen in this way as a matter of course. Furthermore, this, for Anderson, is the reason why cinema often seems real (p.89). Applying this to sound in the FPS game, the player is a part of an acoustic ecology because the player listens with the ears of an ecological organism (a participant in an ecology that is) and, additionally, Anderson's argument applied to the FPS game provides strong support for player immersion within and participation in the game's acoustic ecology because the player responds to it as if it were real, a part of the natural world.

In the natural world, the singing of birds, the sound of a computer fan, muffled conversation in the distance and the drone of an aeroplane all create a continuously evolving acoustic ecology from which I derive meaning and contextualization and to which I contribute with, for example, the rhythmic tap of fingers on my computer keyboard. There are some differences though. I cannot (yet) physically enter the visual world of the FPS game in the same sense that I am physically a part of the real world. More importantly though, my real-world environment is normally a safe environment where I can comfortably push most sound to the background. My auditory system can operate in standby mode (or, in cognitive terminology, my auditory system is operating at a low level of perceptual readiness) awaiting more urgent signals as categorized by experience. I suggest that the hostile world of the FPS game requires a high level of perceptual readiness in regard to sound.

In order to explore the hypothesis, some assumptions are made about the framework through which the FPS game acoustic ecology is created. It is assumed that there is a player, a soundscape (a set of dynamic sounds which the player hears) and a game engine and that these comprise a triangular framework representing the relationships between them. The player is required in order not only to play the game but also to hear the sounds of the soundscape. The player's haptic input[14] is used by the game engine to provide most of the sounds which will be heard in the soundscape. The game engine also provides sounds of its own which may be certain ambient sounds, depending upon which game level is being played, or game status sounds dependent upon the configuration of the game. This game engine comprises the game's computer code and all the peripheral files such as game level files, 3-dimensional models, digital images and, importantly in this case, digital sound files or audio samples.[15] This triangular framework is shown in *Figure 1.1* and is

14 Typically through the agency of a keyboard and mouse combination for an FPS game played on a personal computer (as per most of the examples I provide).

15 In a multiplayer game, a game server is the network interface for all the players' individual game engines.

used as the basis for more complex models as the hypothesis is investigated in the following chapters leading, ultimately, to a complete, diagrammatic model of the FPS game acoustic ecology in chapter 8.

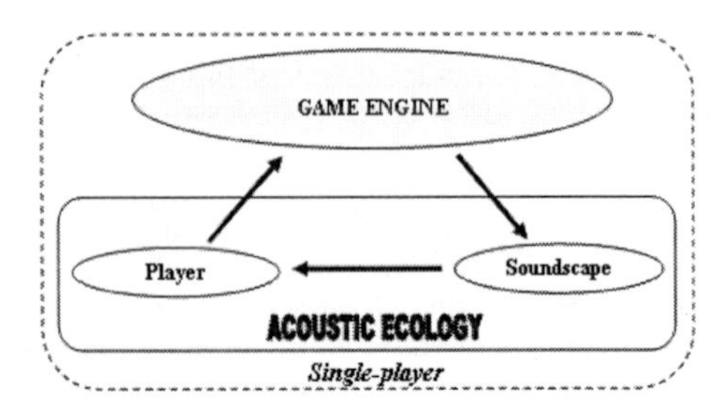

Figure 1.1. Proposed basic model for a framework of the FPS game acoustic ecology (single player).

In order to theorize the hypothesis, a conceptual framework is devised comprising a conceptual language and a taxonomy which builds upon the basic, assumed model in *Figure 1.1* (the methodological approach taken in constructing this framework is described in section 1.6). It is not intended to be a complete explanation of digital game sound and the experience of it but is intended to begin to account for the phenomenon of the acoustic ecology as found in FPS games. Future research, it is hoped, will build upon this framework in order to account for similar or different aural phenomena in other genres of digital game.

1.3 Contextualizing the work within Game Studies

As befits a relatively modern industry, (Digital) Game Studies is a recent creation of academia although it is not quite as young as Aarseth (2001) claims in the following interesting statement:

> 01-02 may also be the academic year when regular graduate programs in computer game studies are offered for the first time in universities. And it might be the first time scholars and academics take computer games seriously, as a cultural field whose value is hard to overestimate.[16]

Although there is a wide variety of approaches to the study of games within Game Studies (both in possibility and in actuality), a quick perusal of some of the more academic game journals such as *Game Studies* (http://www.gamestudies.org) (in particular its first issue), indicates that there is a division (in some cases quite heated) between narratologists and ludologists. This broad statement (and there are

16 As noted below, I wrote the BSc(Hons) Computer and Videogames degree at the University of Salford in England which had its first student intake in 1999. In the United Kingdom, a Honours degree is usually seen as an undergraduate degree whereas in other countries, such as New Zealand, it is a graduate degree. Regardless of this distinction, it should also be noted that the University of Abertay Dundee in Scotland offered a Masters of Science in Computer Games Technology in 1997.

those who attempt to bridge the gap) is perhaps better illustrated by the suggestion that there are those who study games as a form of something else (literature or cinema for example) and there are those who study games as games. As argued in this work, narrative plays little if any part in the FPS game (a tacit recognition of this may be the reason that there is no narratological approach available that researches FPS games or uses them as case studies[17]). As Eskelinen (2001), arguing against the narratologists, so delightfully puts it:

> Games are seen as interactive narratives, procedural stories or remediated cinema. On top of everything else, such definitions, despite being successful in terms of influence or funding, are conceptually weak and ill-grounded, as they are usually derived from a very limited knowledge of mere mainstream drama or outdated literary theory, or both. Consequently, the seriously and hilariously obsolete presuppositions of Aristotelian drama, commedia dell'arte, Victorian novels, and Proppian folklore continue to dominate the scene.

Whilst this work's general approach is one that studies the FPS game as game (rather than narrative), ideas are made use of both from within and without Game Studies. Thus, the work discusses immersion in virtual environments and player experiences in addition to concepts from disciplines that, so far, have not yet made an appearance in the Games Studies debate; disciplines such as sonification studies, autopoiesis and psychoacoustics, for example.[18] This is because firstly, sound in the digital game very rarely rates a mention in the Game Studies literature and, secondly, there are aspects of digital game sound that may not be explained within the range of theories currently found in the Game Studies discipline alone.

In many ways, the academic study of digital games has taken a course that mirrors that of Film Studies both in the areas that are studied (audiences, violence, narrative, spectacle, film as film, game as game for example) but also in its dealings with sound. From its inception, Film Studies made little mention of film sound and one has to fast forward several decades from the birth of the cinema to begin to notice an increasing number of writings on cinematic sound culminating in the devotion of entire books on the subject by the likes of Altman (1992a) and Chion (1994) among others.

The academic study of sound in digital games is likewise currently relegated to the periphery.[19] To provide technical and historical examples mirroring the development of 'talkies', the early computer games (dating from the late 1950's) were silent and the first video game to feature sound (albeit primatively) was *Pong* (Atari, 1972). As with cinema, sound in digital games was developed, on the production side, with improvements in the digital synthesis and recording of sound married to advances in

---

[17] Conversely, in order to advocate the role narrative plays in digital games, it is a good idea to choose the type of game wisely. One reason why RPGs are widely used for this.

[18] And, despite Eskelinen's imprecations, I do also make use of some of the laws of Aristotelian drama in chapter 9.

[19] There are several reasons for this. Like early cinema, early computer games (as a medium) were silent and academic publishing is, in the main, a paper-based medium with little possibility for aural illustrations (unlike images). Furthermore, in the normal course of events, with sound being omnidirectional and no ability to shut the ears (unlike the eyes), sound is often pushed to the background and image is prioritized. Image requires a focal point and thereby provides a focus for discussion.

consumer electronics such as increased computer processing power and memory and a proliferation of audio channels beyond just the one. Whereas the arcade machine on which *Pong* was played utilized one beep only, the XBox of 2001 boasts the capacity for 256 simultaneous 2D sounds and 64 simultaneous 3D sounds[20] (that is, a sound that has been processed for positional and locational purposes) in addition to five digital signal processors (DSPs) two of which are capable of real-time processing of sound.[21]

During the course of this research, very few articles focussing on the player experience of sound in the digital game were found in the Game Studies literature.[22] Furthermore, those articles of relevance which were discovered are short and inconclusive. Sound, therefore, is largely overlooked in the Game Studies literature and it is this lacuna which the research presented here intends to start filling.

In the conclusion to the work, ways are suggested in which this research may contribute to other theoretical understandings of digital games. The study of immersion in game worlds, player experiences and pleasure or navigation through the game environment for example. My personal experience in playing FPS games reflects an engagement with sound on several levels. I can take pleasure in the world of sound in which I am immersed, I can use sound to navigate my way around a game level with which I am unfamiliar, I can use sound to contextualize myself not only within the game world but also in relation to other characters and the gameplay and I can use sound to contextualize other characters and game objects in relation to myself. Because most of these points of engagement with FPS sound concern immersion or presence within the game world it is all the more surprising that those writing about immersion in Game Studies make little mention of sound.[23] It is not possible physically to be present within the game world but this work makes the case that it is possible to be physically immersed within (and contributing to) the acoustic ecology and this possibility demands that greater attention be paid to the role of sound in digital games.

Before providing a methodology and work structure, the FPS game is first briefly introduced as a genre, explaining why it is a fruitful area of research as regards the hypothesis and a short case is made (to be expanded upon in later chapters) for the importance of sound in this genre.

1.4 The FPS game

According to the *Wikipedia* "[a] first-person shooter (FPS) is a combat computer or video game genre, which is characterized by the player's on-screen view of the game simulating that of the character or First-Person view" (2002-2006). *Spasim* (1974) is

---

[20] These are potential internal channels; as the game is being played they will be mixed down to stereo or 5.1 surround sound depending on the external audio hardware being used.

[21] For a more complete history of computer game audio from the technical perspective, see (Collins, n.d.).

[22] There is a larger proportion of articles and even complete books discussing digital game music composition, the game sound business or the technical parameters of game audio hardware for example. These are either non-academic writings or do not concern my hypothesis.

[23] Whalen (2004) discusses the use of the musical soundtrack in digital games to increase or lessen perceptions of immersion but there is no mention of diegetic sounds and Manninen (2003) suggests that non-verbal audio has an effect on immersion in a paper that lists, with little further development, all possible factors influencing interactivity with games.

probably the first FPS game although it lacked many of the features that are taken for granted in a modern FPS game; it had wire-frame graphics and no hand or weaponry on screen for example but was capable of being played in multiplayer networked mode with a first-person perspective. The first FPS game to feature the archetypal hand projecting into the screen was *Catacomb 3D* (Gamer's Edge, 1992) but the FPS game that popularly defined the genre was *Doom* (id Software, 1993).

Although the *Doom* character had limited movement within the game world and there were limits to the spatial design of the game's levels, it features the hands with a weapon perspectivally receding into the screen, texture-mapped surfaces, a range of both claustrophobic and open spaces, multiplayer ability and, of course, a focus on action in the form of killing either bots or player-characters participating via a LAN or through the Internet.

The modern FPS game retains these features but also gives the player's character a relatively large degree of freedom of movement within the game environment[24] and, depending on the sophistication of the game engine, not only localizable sound but, additionally, real-time DSP of audio samples to match the dimensions and materials of the visual environment presented on screen. Examples include *Quake III Arena* (id Software, 1999), *Half-Life 2* (Valve Software, 2004) and *Doom 3*. Many FPS games give the player the opportunity to design his own levels and some game engines, such as the one used in *Quake III Arena*, are licensed free for non-commercial use. This practice has given rise to the phenomenon of 'modding' whereby fans can adapt the game engine to create new FPS games. For example, *Counter-Strike* (Valve Software, 1999), deemed by *Wikipedia* to be the most popular multiplayer FPS game today (2002-2006), was originally a mod of the single-player FPS game *Half-Life* (Valve Software, 1998). Modifying the game engine requires knowledge of computer programming (usually C or C++). However, many FPS games also have level editors (either first-party or third-party provided) to allow the player to design his own levels within the specific FPS game — such levels not only being structurally designed by the player but having the ability to use the player's own graphics and sounds over and above the developer-supplied resources — and that do not require any computer programming knowledge.

Games such as the *Doom*, *Quake* or *Half-Life* series are a particular subgenre of the FPS game known as the FPS run and gun game. These FPS games typically have a focus on violent action, in the form of killing other game characters, at the expense of long-term strategy or complex narrative. Which is not to say that strategy and narrative play no part — *Half-Life 2* (Valve Software, 2004) is a case in point requiring the player to strategize his gameplay to reach an ultimate goal over many hours of gameplay as opposed to *Quake III Arena* whose purpose is simply to win each level with short bursts of frenetic, visceral activity, such levels being independent of each other in terms of objectives and strategies. Furthermore, the player is encouraged to believe that he is immersed within an illusory 3-dimensional game world and he is, in fact, able to interact with it in significant ways. The hunter and the hunted premise of these games (and therefore the need to attend to sound cues for the survival of the player's character) combined with the possibilities for

---

24 There are limits to this movement which are dictated either by the game engine's physics (such as the restrictive effects of a simulated gravity) or by the level design which may, for example, feature images of doors which cannot be opened or walls which cannot be climbed over. Much of this is designed to lead the player along pathways to particular objectives or action hotspots within the game.

immersion in and interaction with the game world makes such game types appropriate to the study of the acoustic ecology of the game. Hence the focus of this work on FPS run and gun games and the reader may assume that further mention in this work of the term *FPS game* always refers to the modern FPS run and gun game unless otherwise stated.

All FPS games, as the name implies, present the player with a first-person perspective where the image presented on screen is intended to be what the player would see were he physically to take the place of his character within the game.[25] The more modern FPS games have increasingly detailed and sophisticated visual and sonic environments making use of perspective, scaling and parallax techniques in addition to multiple audio channnels, real-time mixing and localizable sound among other features. Additionally, continuing the tradition started by *Catacomb 3D* (Gamer's Edge, 1992), modern FPS games enhance this first-person perspective by the addition of a hand or pair of hands on the screen clutching a weapon and that recedes perspectivally from the player into the game world. The fact that these hands typically respond to player input with a variety of animations (such as reloading a gun at the player's command) further enhances the desired immersion into the game world through identification with the player's character.

Many FPS games do, though, offer the possibility to play in third-person perspective where the player's line of sight is usually positioned behind and above the character on screen. An example of this is *Unreal Tournament 2004* (Epic Games & Digital Extremes, 2004), in which the player can play the entire game in third-person perspective if they wish, or *Halo* (Bungie Studios, 2001-2004) in which the player is automatically put into third-person perspective upon entering a vehicle). Additionally, third-person perspectives are sometimes offered in a variety of in-game scenarios such as an overhead view of the player's character lying dead on the ground while waiting to respawn[26] or in allowing the player to spectate by following another bot or player's character within the game (although this latter case means that the player is not taking part in the game and any other players are unaware of such spectators). The main mode of play though, and the one for which the FPS game is designed and marketed, is first-person perspective and this is the mode that is focussed upon in this work.

FPS games come in single-player or multiplayer networked configurations, or both, allowing the player to choose which configuration to play in after launching the game software. Single-player games allow for one player to use his weaponry against computer-generated characters (bots) which are programmed to either attack the player or, in team-based gameplay, to co-operate with the player towards a common goal (such as killing all the enemy team's bots). Bots make use of increasingly sophisticated artificial intelligence (AI) routines in order to seek out and attack the player, to dodge the player's advances or to work as a member of one team competing against another team.

With increases in computer network bandwidths and the growth of the Internet,

---

[25] Albeit with some limitations such as a lack of peripheral vision. Furthermore, framed as it is within the boundaries of a visual monitor, the image on screen competes with the images from the player's environment.

[26] Following the character's death, the act of respawning (usually carried out automatically by the game engine after a short period) re-initializes the player's character at one of possibly several spawn points in order to join the game once more.

almost all modern FPS games provide both single-player and multiplayer configurations where, in the latter case, networked players compete against, or play as a team member with, other players connected to a central game server via either a Local Area Network (LAN) or through the Internet.[27] There are a variety of game modes variously offered by different FPS games including the free-for-all or deathmatch mode (in which players attempt to kill as many other players' characters as possible) and team-based modes such as team deathmatch and capture the flag (in which points are gained by assaulting the enemy team's stronghold, capturing their flag and transporting it safely back to the player's base). This work concentrates initially on the acoustic ecology of the single-player FPS game but, by chapter 7, expands the conceptual framework to include multiplayer games.

The FPS game affords the player the perception of moving around the game world. This illusion is achieved by the game code creating the game's visual environment in real-time, by (re)positioning the game's visual objects (affording the illusion of parallax), by scaling visual objects differently (affording the illusion of visual depth) and by moving sound objects around the player as he issues movement commands. This is one of the ways in which the FPS game attempts to immerse the player within the virtual space of the game world but has been critiqued by Laurie Taylor (2003):

> The very attempt to bring a player into the game space through the screen by means of a first-person point-of-view is, ironically, inconsistent because the first-person point-of-view assumes that the player himself can be caught into the structure of the game and can then be incorporated into the game space. In this way first-person perspective assumes that by enveloping the player as the player into the game space, the player becomes part of the structure of the game space (p.7).

Immersion in the FPS game acoustic ecology is a common thread throughout the work; in particular, it is argued that the player, if he cannot become physically immersed in the game space by visual means, is, in fact, immersed in the acoustic space of the game.

Having defined the FPS genre, and stated that this work concentrates on the modern FPS run and gun subgenre, the following section makes the case that this genre is a particularly good entry point for any discussion of digital game sound because it has the potential to engage and immerse the player within the gaming experience and the game world, more so, perhaps, than any other commonly played digital game genre.

1.5 Sound in the FPS game

Articles or even complete books on Game Studies have been, at best, woefully coy about sound in the game environment or have, at worst and paraphrasing Chion's words, completely ignored it (p.xxv). Statements such as "the only thing the player knows about the world of the game is what is displayed on the screen" (Kücklich, 2003, p.5) may be valid for certain types of digital games,[28] but ignore the important

---

[27] A LAN is a small network of hosts (computers, network devices). Several LANs may be connected together in a Wide Area Network (WAN) and the Internet is a network of WANs.

[28] And even this, at first sight, is doubtful.

role that sound plays in many digital game genres not least of which is the FPS game. Although discussed throughout this work, it is enough here to note some simple examples demonstrating the utility of sound to the FPS game: that the game code will often play certain sounds as a signal that something significant has happened in the gameplay (such as the capture of a flag) in an area of the game world which may not be displayed on screen; that many FPS games utilize a system of radio messages (recordings of speech) as a form of communication between team members dispersed throughout the game world and that a player is able to have some awareness (and often this will be the first such awareness) of what is happening in the vicinity, but off-screen, through the agency of sounds like approaching footsteps or weapons fire.

The FPS game is a hostile environment where, if the team is to win and the player's character to survive, the player must be attentive to (and learn the meaning of) all possible sonic cues. Rather than operating in standby mode, the player's auditory system must be on full alert. Furthermore, it is my contention that sound is the most important contributing factor to the creation of the 3-dimensionality of the game world and, therefore, the immersion of the player in that game world. The image of the game environment on a 2-dimensional screen must fake 3-dimensionality; sound is (always) 3-dimensional and, additionally, has the potential to convey information about the virtual materials and dimensions so artfully represented in the image.

There are two basic functions that sound provides in the FPS game: feedback to the player and, related to this, an awareness of operating within, and being a constructive component of, a space or spaces within the game. The scenario from an FPS game played out in the Prelude above would be very different without sound and our hero would have difficulty contextualizing himself, immersing himself within the game world and making sense of the game's action were the game to be entirely mute. Without sound, it truly would be the case that the only knowledge which the player has of the game world and the action (barring the use of prior knowledge) derives from what is shown on the screen as Kücklich suggests (p.5). Furthermore, not only would he find it more difficult, if not impossible to locate friend and foe, but, without sound, they would experience similar difficulties in locating him. Finally, the sounds of unseen actions can provide not only an objective response (that is, for example, *there is action taking place over there and how do I, as a team member, use that to contribute to my team's victory?*) but also a subjective, emotional response (and, at times, a physical, whole-body response) that is intimately related to the continually morphing sonic environment of the game. The addition of sound to the FPS game not only opens out the world of the game from the restricted 2-dimensional image displayed on the screen to a full 360º space but also adds other levels of sensory and perceptual experience and aids in a contextualization within the game's action.

What is sound in the FPS game? This might seem a question in search of an obvious answer. *Res ipsa loquitor*: digital game sound is that which the player hears and that is provided by the game code. However, Stockburger (2003) differentiates between sound in the *user environment* and sound in the *game environment* as an expanded definition of digital game sound. Whilst acknowledging that sounds from the user environment may have an impact on the player's experience of the game, this work concentrates solely on sound from within the game environment for a number of reasons. The work proposes that the sound of the FPS game is an acoustic ecology and part of the process of the creation and sustainment of this ecology involves player immersion within the game world; any sounds external to that game world act as perceptual shocks potentially breaking the illusion of immersion by reminding the player that he is operating in at least two different environments, one real and one virtual. Additionally, sounds from the user environment are outside the control of game designers and this work has, as one of its secondary objectives, the purpose of providing an increased understanding of how to design the acoustic space of the game, all the better for immersion within the game world, and, therefore, how to create a viable acoustic ecology. Furthermore, most FPS players, as pointed out in chapter 5, will play with headphones in an effort to block out sounds from the user environment and to increase the effect of immersion (Morris, 2002).

Game developers have been quick to exploit increased graphics capabilities to create often stunning virtual, visual environments, but as yet seem leery of exploiting the potential of audio technology other than as a bit-part player to the visual star. This concentration on the visual at the expense of the auditory is paralleled by many games studies authors who for example, when discussing commonalities between film and computer games, lavish considerable effort on visual image but neglect to mention sound (King & Krzywinska, 2002 for example) and likewise when discussing the construction, in film and games, of narrative or other forms of space and place (Liang & Tan, 2002 for example).[29] It is also mirrored in the many Internet forums devoted to discussions of FPS games and in the marketing material within which the game is packaged. Thus, a quick perusal of reviewer and fan comments on third-party levels for *Quake III Arena* at one Internet site (http://lvlworld.com/top.php?v=40) for example, yields little in the way of discussion of sound while the back-of-the-box marketing hype for *Doom 3* stresses the visuals but makes no explicit mention of the sound: "incredible graphics, and revolutionary technology combine to draw you into the most frightening and gripping first person gaming experience ever created". Advertising for FPS games is primarily a visual medium that uses screenshots of game levels on Internet sites and pictures of characters and visual environments from the game on the game box[30] — there are no sound grabs from the game on offer.

The FPS game is a fruitful area of research into digital game sound for several reasons. In few other types of digital game is the ability to auditorily perceive the virtual world and to pro-actively react to sound of such importance to gameplay and the gaming experience than in the FPS game. The FPS game typically provides a first-person perspective which is designed to place (or at least give the illusion of placing) the player within the virtual world as presented on the screen and, in part, created through the sound. This, in turn, prioritizes the aforementioned hunter and the hunted scenario (in combination with the player's abilities to attack and defend) and the enhanced perception of immersion within the game world.

The game system has not been designed yet that allows the player to physically inhabit the game architecture and nor, when that time comes, would the human

---

[29] This thesis concerns itself with audiovisual games, although it is hoped it will be of some use to those readers interested in the expanding repertoire of sound-only games; such readers may be directed to *AudioGames.net* (http://www.audiogames.net/) as a starting point or to the writings of those authors specifically dealing with that field (for example McCrindle & Symons, 2000; Röber & Masuch, 2005; Sánchez & Lumbreras, 1999; Velleman, van Tol, Huiberts, & Verwey, 2004).

[30] In some cases, with an exaggerated detail and realism that is disappointingly not replicated within the game.

visual system permit an omnidirectional view of that architecture. Sound, especially when presented through headphones, is very different. It is an omnidirectional experience, carrying information about the game environment and game action both on-screen and off-screen, information about the sound object's source, and emotional or mood-inducing information. Furthermore, FPS game sound can act as an audio beacon guiding or encouraging the player to particular points within the game level and game sound informs about objects in the game which, although they may be in the player's line of sight, are hidden from view by other objects on-screen. Finally, the nature of FPS games is such that the player is placed in the position of being the hunter or the hunted where both visual and auditory systems must be on the highest alert for, or, in perceptual terms, must give the greatest attention to, the slightest of cues or affordances from the game environment. By attending to and reacting to sound cues from the game environment, the player has the power to significantly redirect the gameplay; being able to respond to heard sounds and contributing sounds oneself is one of the criteria for participation in an acoustic ecology. Few other game types have such auditory potential and it is for this reason that this work will focus its efforts on the FPS game (that is, the FPS run and gun game), using it as a model to produce a framework for the study of digital game sound (and such a methodology may help the analysis of sound in other digital game types by a process of exclusion).

The work is concerned solely with gameplay sound, that is, the sound that players hear while playing the game. While I distinguish between such gameplay sounds and the interface sounds which may be heard when the player is navigating game option menus, either between game levels or while the gameplay is paused, I do not include the latter set of sounds in the analysis because I argue that they do not form part of the acoustic ecology and indeed act to perceptually extract the player from any acoustic ecology which the gameplay constructs. In this way, they act as perceptual shocks reducing the sense of immersion in the game world. Neither do I concern myself with any nondiegetic music which may be heard during gameplay other than to note it as a point of difference to diegetic sounds. The volume of such music can usually be adjusted separately to the volume of all other sounds in the game and many FPS players, myself included, prefer to turn it completely off because it is a distraction and often makes gameplay sounds difficult to hear.[31] Arguably such music may be viewed as part of the acoustic ecology that is created during gameplay but the analysis of it, and any meaning derived from this music by the player, requires a very different methodology and research technique than is required for the study of other gameplay sounds. Such an analysis is for future research that builds upon and adds to the findings of this work.

Among the variety of academic disciplines that study sound there are a number of approaches using each of the areas of sensation, perception and cognition of sound either in isolation or in combination. For example, the science of acoustics is in large part based on the study of the sensation of sound and the physical attributes of sound (frequency, amplitude and time — referred to by Gaver (1986) as the dimensions of sound), psychoacoustics bases its approach mainly on perception of sound and psychology concentrates on cognition of sound. This is not to say that no acoustician investigates sound from the point of perception or that no psychologist uses the scientific study of sound as sensation, a phenomenological approach when

31 I do sometimes make an exception to this as noted in chapter 7.

enquiring into the meaning of sound, but there is a general compartmentalization of sonic study that has developed since the nineteenth century through the twentieth century. As Gibson (1966) remarks in an objection to the use of physical acoustics (the study of sound as sensation) alone to study sound: "It treats physical sound as a phenomenon *sui generis*, instead of as a phenomenon that specifies the course of an ecological event; sound as pure physics, instead of sound as potential stimulus information" (p.86).

However, since the latter half of the twentieth century there has been a realization that a thorough study of sound requires knowledge of each of sound sensation, sound perception and sound cognition and this idea has been advanced in part by the work of authors such as Gibson and Schafer (1994) while authors such as Gaver (1986) argue for a study based on the dimensions and properties of the sound source rather than dimensions and properties of sound. It is probably the case that the contemporaneously increasing facility with which sound could be recorded, edited and processed (the development of magnetic tape recorders and synthesizers for example) played a part in this realization. In some cases, this cross-fertilization has led, in part, to the development of new technologies, such as the perceptual encoding of audio as enabled by MP3 algorithms, and in others it has led to new areas of study, such as sonification and the design of auditory icons not to mention ecological acoustics.

It is important therefore, that a broad theoretical approach is taken for this research, an approach that covers not only a phenomenological stance but also includes the perceptual and cognitive point of view that helps explain how players experience sound within the game. The following section, then, outlines the methodology of the work.

1.6 Methodology and organization of research

This section details the methodology of the work and the practical organization of the research entailed in that methodology.

1.6.1 Methodology

In attempting to lay the foundations for digital game sound research, this work takes a very catholic, multidisciplinary approach. Because so little has been written about digital game sound as it relates to the broad boundaries of the hypothesis (that the sound heard in an FPS game may be construed as an acoustic ecology of which the player is an integral and contributing component), the work draws upon concepts derived from other fields of research in order to construct a conceptual framework for the explication of the hypothesis of the FPS game acoustic ecology. The work attempts to position itself at the forefront of digital game sound research by providing the concepts that are needed to discuss sound in this context. It is not possible, for instance, to talk about the economic history of digital game sound or its political ideology if one has no sonic language within which to frame such a discourse. Such future research, it is hoped, will build upon the conceptual framework formulated here.

Consequently, the work's methodology comprises the formulation of a conceptual framework to be used primarily for the exposition of the hypothesis but which may also be used for future research into the wider area of digital game sound. This framework draws on concepts from a wide range of disciplines and is a selective process involving a large field of potentially relevant ideas about sound. Only those that have the capacity to contribute to the hypothesis will be considered in depth. Other ideas will be treated in lesser detail if, whilst not being able to contribute to the notion of the FPS acoustic ecology, they nevertheless prove to have some worth in the construction of the conceptual framework. No one discipline has been chosen precisely because little research has so far been undertaken into digital game sound (there is, therefore, no dogmatic stronghold worthwhile investing from one theoretical opinion or another) and because "[e]xtending one theory too far, into a domain for which it was never meant, does no one a service" (Kress, 2003, p.13). The methodology, therefore, is aimed at researching and adapting concepts from a range of disciplines and, where required, proposing new concepts. Combining these concepts, adapting them for the medium of the FPS game and formulating new concepts leads to the construction of the conceptual framework that is ultimately used to explicate the hypothesis.

Section 1.2 proposed that the acoustic ecology of the FPS game comprises, at a fundamental level, a player and a soundscape and the relationship between them. This proposition raises some questions about the roles of these components in the acoustic ecology and the research required to answer these questions will help in establishing the conceptual framework.

Perhaps the fundamental question to be asked is *How is meaning derived from sound?* To answer this in the context of the FPS game, one must investigate not only sound as sensation (sound *sui generis*) but also the perception and cognition of sound. Furthermore, as sound is *designed* for the FPS game, it may be assumed that it is designed with a particular intent in mind that, ideally, will be immediately understood by the player or, with increasing experience in the game, may be learned. Thus, disciplines dealing with the design of sound, the encoding of meaning in audio data and the semantics of sound form part of the research.

The examples of FPS run and gun games used throughout the work are audio-visual games as opposed to audio-only games. Consequently, a question to be asked is *What is the relationship between sound and image in the FPS game?* Although image is not a part of the acoustic ecology, the FPS player may navigate within the visual game world displayed on the screen and this dynamic relationship between the player's character and other objects in the game changes the soundscape that the player experiences. There is, therefore, a relationship between sound and image that influences the acoustic ecology and disciplines that describe and analyze this relationship in both digital games and other media will be useful to the research.

Because I suggest in section 2.1 that the player is a fundamental component of the FPS game acoustic ecology, the question needs to be asked *What is the role of sound in immersing the player in the acoustic ecology?*[32] This question poses others pertaining to the nature of the acoustic spaces in which the player may be immersed. As a consequence, the research must assess disciplines covering the perception of spaces and immersion in virtual environments for concepts which may be used or adapted to form a part of the conceptual framework as it relates to spaces and

[32] There are many other aspects describing player immersion in the FPS game world but this thesis deals solely with immersion through and in sound.

immersion in the FPS game acoustic ecology.

The work initially works towards the construction of a conceptual framework for the explication of a hypothetical acoustic ecology comprising just one player and one soundscape. This is related to the FPS game played in single-player mode. However, many modern FPS games (including all the examples I use to illustrate the work) may be played in multiplayer mode and therefore, I suggest, involve multiple acoustic ecologies. In this scenario, players may perceive the actions of other players on the basis of sounds triggered in the game. Thus, a further question is *What are the relationships between players on the basis of the sounds they trigger?* To help construct a conceptual framework which accounts for these relationships, a variety of relevant disciplines will be researched.

The methodology, then, comprises a review of a range of disciplines in order to extract and adapt concepts for use in the conceptual framework. The goal is to construct a conceptual framework whose primary purpose is to elucidate the hypothesis but which is flexible and broad enough to have potential future use in investigating sound in genres of digital games other than FPS. The disciplines chosen or adapted, and the selection of concepts from these disciplines, proceeds, therefore, on this basis.

The selected disciplines are researched in the central chapters of the work as presented in section 1.7 below that details the structure of the work. Preceded by an initial chapter that discusses current Game Studies research on digital game sound, these chapters are: 'Meaning in Sound'; 'Sound, Image and Event'; 'Acoustic Space' and 'Diegesis and Immersion'. For the most part, I have tried to maintain discussions central to established disciplines in the one chapter. However, this is not always appropriate and so, where necessary, some themes from a specific discipline may appear in chapters other than where that discipline is more fully discussed. For example, film sound theory appears mainly in chapter 5 but a discussion of diegesis (as debated in film sound theory) occurs in chapter 7 because diegesis, as I define and adapt its use (in film sound theory) to the FPS game acoustic ecology, is intimately connected to immersion in that acoustic ecology.

1.6.2 Organization of research

In order to manage the research taken from the disciplines which may be of relevance to the hypothesis, I have made use of self-authored, bespoke software that allows me to rapidly sift through references, quotations and comments and that permits me, in practical terms, to write this work.

As part of the practical methodology used to implement this work, I have developed and made use of a software tool called *WIKINDX* (Grimshaw, 2003-2006). Developed under the *GNU General Public License* (GPL) (Free Software Foundation, 1989), *WIKINDX* is a web-based, database-driven tool written in PHP to manage bibliographies and metadata[33] and is accessible using any graphical web browser on a networked computer.[34] The decision was taken to develop this software because I needed a way to store and to sort rapidly through all my research references and metadata in a manner and with an efficiency that index cards or existing commercial

[33] Quotations, paraphrases, ideas and comments.

[34] *WIKINDX* may also be installed in single-user mode on a stand-alone, non-networked computer that has been installed with the appropriate support software.

reference management systems do not allow. Over the span of three years, this initial brief expanded so that is it now possible to write entire, fully-formatted and accurately cited articles (where such citations are context-sensitive) within the one software.

Although a full list of the main features can be read at the project website, there are two functions that have dramatically eased the research and writing of this work. The first is the ability to rapidly search the metadata in the database by a variety of search parameters such as search word(s), author name(s) or keyword(s). These metadata may also be cross-referenced to other metadata or bibliographic references enabling a web of knowledge spanning diverse fields of interest to be constructed. This allows a rapid harvesting of research material organized by themes when it comes to writing a research paper.

The second major function in *WIKINDX* is its integrated WYSIWYG[35] word processor. This allows the user to write publication-ready research papers (or PhD work chapters) from within the web browser importing citations and metadata direct into the text and applying a range of text and font formatting. The paper may then be exported to Rich Text Format (RTF) allowing its opening in most commercial word processors and, if necessary, the compilation of such papers into a work or a book. With this export, the user may format the citations and append a bibliography in one of several in-text or footnote-type bibliographic and citation styles.[36]

Because *WIKINDX* is free software and is released under the GPL, its code is available for other similar projects to use and major sections of the code have been extracted as separate downloads which form the basis of another GPL project called *Bibliophile* (http://bibliophile.sourceforge.net/). This is a loose grouping of free bibliographic software projects. At time of writing, the following bibliographic software is known to make use of code originally developed for *WIKINDX* (in particular the parser for bibTeX files and the bibliographic/citation formatting engine (OSBib)): *Aigaion* (2004-2006), *Bibliograph* (Boulanger, 2006), *BibORB* (Gardey, 2005), *PHPBibMan* (Pozzi, 2004), *PHP Bibtex Database Manager* (2006), *refbase* (Steffens & Karnesky, 2006) and a dokuwiki plugin (Ambroise, 2005-2006) has been developed using OSBib. Furthermore, because *WIKINDX* supports multiple users and because its databases can be accessed through the World Wide Web, there are several publicly available wikindices run by various institutions and organizations including Princeton University (http://www.kuyperresearch.org/wikindx3/), Wikimedia Germany (http://tools.wikimedia.de/~voj/bibliography/) and NASA (http://outtek.grc.nasa.gov/Soils-Bib/).

This whole work has been organized and written within *WIKINDX*. The bibliography and relevant metadata gathered for the research is publicly available at the *Digital Game Audio WIKINDX* (http://sad-web.wlv.ac.uk/~mark/wikindx3/) and is the core from which future publications will be fashioned.

---

35 What You See Is What You Get.

36 *WIKINDX* has *American Psychological Association*, *British Medical Journal*, *Chicago*, *Harvard*, *Institute of Electrical and Electronics Engineers*, *Modern Language Association* and *Turabian* and includes a style editor for the user to create her own bibliographic and citation formats.

1.7 The structure of the work

This work is divided into three main sections: chapter 1 'Introduction' (this chapter), six chapters detailing the research undertaken into a variety of disciplines and that work towards constructing a conceptual framework for the analysis of the FPS game acoustic ecology, followed by a chapter summarizing the conceptual framework, and the concluding two chapters. There is a narrowing down of the work topic from broad outlines of what FPS game sound is and the possible ways of interpreting it in the initial chapters to a focus on the conceptual framework and the hypothesis in the final chapters. Additionally, an appendix describing *Urban Terror* and listing its audio samples and an appendix comprising a glossary of important terms appear towards the end of the work.

Chapter 2 ('Current Research on Digital Game Sound') is a literature review of current writings focussing on the player experience of digital game sound and that attempt taxonomies of that sound. The next five chapters not only present potentially useful ideas from a range of disciplines but also, in the partial construction of the conceptual framework that they contain, make use of and develop what has been discussed in each preceding chapter. Chapter 3 ('Sound Design and Production') deals with technological issues as they relate to the hypothesis, in particular the mode of production of sound at the game development stage and the organization of sound in the distribution medium and its use by the game engine. Chapter 4 ('Meaning in Sound') provides much of the conceptual background for the following chapters in that it describes how listeners derive meaning from sound (using the terminology of cognitive science). Two disciplines, sonification and auditory icon design, are presented not only to illustrate the human understanding of sound and some of its practical applications but also because they have the potential to explain the experience of sound in the FPS game — both are explicitly concerned with sound and computer systems, with designing meaning into sound and with deriving meaning from sound. Chapter 5 ('Sound, Image and Event') investigates the understanding of sound in the context of image. Although ideas from other research areas are debated, in large part this chapter makes use of theoretical writings on film sound. A discussion of relevant spaces in the FPS game and the role of sound in the creation of the perception of such spaces is the substance of chapter 6 ('Acoustic Space') and chapter 7 ('Diegesis and Immersion') debates the role of sound in player immersion in the game and understanding of the game diegesis in an attempt to define relationships between players and between players and the FPS game engine as they are facilitated or experienced through the medium of sound.

Chapter 8 ('The Conceptual Framework') is a bridging chapter between the extended literature review and research of the preceding chapters and the conceptual approach of the penultimate chapter. It summarizes and consolidates the conceptual framework, illustrates it with examples drawn from *Urban Terror* and, ultimately, presents a complete diagrammatic model of the FPS game acoustic ecology. Chapter 9 ('The Acoustic Ecology of the FPS Game') discusses and describes the FPS game acoustic ecology in its entirety before proceeding to conceptualize it as an autopoietic system in order to account for the multiplayer FPS game. Additionally, chapter 9 presents a discussion of the FPS game acoustic ecology as dramatic performance as a means to test the robustness and versatility of the conceptual framework and model. Finally, chapter 10 ('Conclusion') presents the conclusions of the work, its contribution to Game Studies and other fields, discusses its strengths and weaknesses and provides direction for future research building upon the discoveries made here.

1.8 Personal background as it relates to the research

Copier (2003) makes the point that the background and experiences of games researchers should be taken into account when engaging upon research whether these experiences are related to games or not. Accordingly, this brief section makes the case for my research on more personal grounds.

Not only would I consider myself an FPS gamer (almost exclusively so, although I no longer play the several hours a day that I used to) but I have had a professional and academic interest in sound for all of my working life. In addition to a period as a sound engineer in Italy, I have an Honours degree in music (majoring in electro-acoustic composition) and a Masters of Science in music technology. Prior to arriving in New Zealand, I worked ten years at the University of Salford in England initially teaching music technology and sound recording, acoustics and psychoacoustics in addition to other topics such as African music and web design. Later, I developed and lectured on the country's first Honours degree devoted entirely to the study of computer games, my teaching being in the areas of level design and game sound design.

As part of this degree, and taking advantage of the university's fat connection to the Internet (part of the *Joint Academic Network* (JANET)), I was able to set up publicly-available game servers devoted to *Quake III Arena* and its variants and, in time, this network of up to 15 game servers became the most popular in Europe.[37] This system further provided me with a public platform for *Quake III Arena* levels I designed and which enabled me to experiment with the role that sound plays in such virtual environments.[38]

It was only natural then, that when it came to choosing a PhD topic that I should choose to marry my twin passions for sound and games to arrive at the topic for this work.

1.9 Summary

In this work, I present the hypothesis that the sound heard when playing an FPS run and gun game may be treated as an acoustic ecology in which the player is an integral and contributing component. The hypothesis, then, is positioned within a larger field of research which concerns itself with the listener's engagement with sound. Like a real-world ecology, most actions result in a sonic consequence which provides information to the participants in this ecology not only about the actions of the participants (themselves and others) but also about the environment in which those participants find themselves. In turn, this, like a real-world ecology, results in change in the aural landscape, the soundscape, and such change potentially

---

37 The web site maintaining these statistics has not been available for several years.

38 Two of these levels, *Grim Shores 2: Devil's Island* (Grimshaw, 2001a) and *Grim Shores 3: Atlantis* (Grimshaw, 2001b), remain amongst the most popular *Quake III Arena* capture the flag downloads at *LvL* (http://lvlworld.com/top.php?m=CTF&v=40)

provokes further sound-producing responses from its participants. That there are relationships between participants, responses from participants and relationships between participants and their environment as a result of sound events is what leads me to propose the view of the sound of the FPS game as an acoustic ecology. The process by which this takes place, then, is the substance of this work.

The sound that concerns this work is the diegetic sound that is heard during gameplay and so does not include nondiegetic music. Neither does the work concern itself with much of the technical nature of digital game sound hardware and software, for which numerous articles exist describing the design of game audio engines, audio hardware for games and descriptions of the audio capabilities of various consoles and computer platforms for example, other than where such a technical exegesis has import to the player's relationship to the acoustic ecology.

The work, then, investigates the role that sound and the player play in the construction of the FPS game acoustic ecology and the changes in that acoustic ecology resulting from interactions and relationships between players and between players and the soundscape. As there is little written on game sound in the Game Studies literature, a variety of related research fields are assessed as to their relevance to the work. This methodology is used to construct a conceptual framework that is then used to explicate the hypothesis of the FPS game acoustic ecology.

Chapter 2

Current Research on Digital Game Sound

Introduction

This chapter marks the start of the derivation of the conceptual framework focussed upon the questions asked in chapter 1 and that forms the basis for exploring the hypothesis of the FPS game acoustic ecology. Because there are so few writings in the field of games studies which deal solely or substantially with digital game sound as it relates to the player experience, and because these writings are similar in their method (that is, most of them are focussed upon deriving taxonomies of digital game sound as a first approach to its study), I initially present a section that summarizes this research, briefly noting points of interest to the hypothesis, before providing a section that more fully discusses its relevance to this work.

To paraphrase Chion (1994), theories of digital games have until now tended to elude the issue of sound, either by completely ignoring it or by relegating it to minor status (p.xxv). Although there is a wealth of information on the practical and technical aspects of game sound design (Bernstein, 1997; Boyd, 2003; Bross, 1997a, 1997b; Zizza, 2000; Turcan & Wasson, 2003; Brandon, 2005 among others), only a handful of writings have appeared that deal explicitly not only with how game sound is designed and organised but also with how players sense, perceive and respond to it (these are discussed in section 2.1). In other words, a theoretical approach to the relationship between digital game sound and the player.

The few writings available which focus on theoretical aspects of digital game sound (as opposed to game music, details of console or computer audio hardware or aesthetic and technical advice on how to create a musical score or sound effects for games) are either conference presentations or are in the form of an internet blog. It is, therefore, not just the paucity of materials on the subject but also the presentation of these materials that reflect the inchoate nature of the study of digital game sound; there are, for example, no book chapters on the subject let alone complete books as exist for film sound (this is still a relatively young industry).

This rudimentary state is further evidenced by the fact that most of these writers devise a taxonomy of digital game sound (as a first approach to analysis) in order to facilitate further theoretical discussion. This derivation of taxonomies is something that is common to both these writings and this work. It is a recognition that the first requirement for a discussion of digital game sound is the development of a conceptual framework which potentially includes a taxonomy. Furthermore, several of these writers present their taxonomies on the basis of case studies or by illustrating them with examples from specific games and this is further evidence of the nascent state of the study of digital game sound; case studies are often used as a first approach to the study of new territory.[1] Such taxonomies (as provided by other

---

[1] Although space precludes a thorough case study here, *Urban Terror* (Silicon Ice, 2005) is used to illustrate the points made in chapters 8 and 9 while examples from a range of other FPS games are used in earlier chapters. A comprehensive case study based on the findings of this thesis is a future task.

writers) go part of the way to explaining what digital game sound is, provide an indication of the particular focus of the writers (for example, which disciplines they approach the subject from) and help to provide starting points in this work for a discussion of the taxonomies' strengths and weaknesses. Those that I evaluate as relevant to a view of FPS game sound as an acoustic ecology I appropriate and develop further in light of ideas from other disciplines. Those that are not relevant (because they do not serve to answer the questions proposed in chapter 1) are discarded.

2.1 Theoretical writings on digital game sound

Stockburger investigates sound in *Metal Gear Solid 2* (*MGS2*) (2001) using a methodology based, in part, upon film sound theory (Stockburger, 2003). In doing so, he applies some of the ideas and terminologies of figures such as Altman and Chion (for example, acousmatic sound and diegetic sound which I discuss further in chapters 5 and 7) to the sounds of the game. For instance, sounds in the game are sound objects and may be acousmatic or visualized. His taxonomy of sound derived from *MGS2* is outlined below in *Table 2.1*:

Speech sound objects	*Effect sound objects*	*Zone sound objects*	*Score sound objects*	*Interface sound objects*
The human voice.	Sound produced by objects within the game environment.	Sounds pertaining to locations or zones within the game environment.	The non-diegetic musical soundtrack.	Non-musical, non-diegetic sounds.

Table 2.1. Stockburger's classification of sound objects in *Metal Gear Solid* (2001).

Stockburger's taxonomy is one based on the organization of sound objects within the game code, the source of sound objects within the game and the means of production of the sound objects (musically composed sound or spoken sounds, for example). Although stating that, in addition to spatial representation, sound functions "to inform and give feedback, to set the mood, rhythm and pace, and to convey the narrative" (Stockburger, 2003, p.1), none of the above classes serve to inform this insight. Furthermore, having discussed Pierre Schaeffer's modes of listening (such as the causal and reduced listening modes that have been adopted by the likes of Chion and which I discuss further in chapter 4),[2] Stockburger makes the useful suggestion that, while such modes may be of use in the analysis of digital game sound, players listen semantically in order to construct relationships between the sound and visual objects in the game (p.4). It is surprising, therefore, that the taxonomy itself above makes reference neither to Schaeffer's modes of listening nor to any semantics of sound.

[2] Briefly, these modes of listening are approaches that listeners adopt in order to analyze some component of a heard sound. Causal listening concentrates on assessing the sound source properties, reduced listening is a concentration on the qualities of the sound itself and the semantic listening mode utilizes (semiotic) codes to analyze specific meanings conveyed by the sound (such as the semantics of speech, music or alarms for example).

Stockburger does, however, provide the important concept of kinaesthetics as a point of difference between digital games and films and suggests that it is "a key factor in creating unique spatial experiences when playing computer games" (p.9) Kinaesthetics relates to the player's ability (in most digital games and certainly in modern FPS games) to turn their character towards the source of an unseen sound. As a concept, it has particular implications for immersion in the FPS game acoustic ecology and I return to it in chapters 5 and 7.

Folmann, discussing sound use in digital games in general, provides four *dimensions* of game audio: *vocalization*; *sound FX*; *ambient FX* and *music* as explained in *Figure 2.1* and *Tables 2.2* and 2.*3* below:

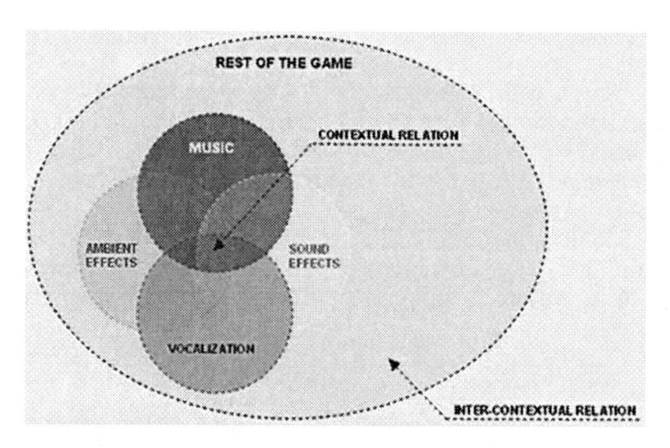

Figure 2.1. The dimensions of game audio according to Folmann (n.d.).

Folmann suggests that these four dimensions not only complement each other but also relate to other aspects of the game by having an effect on the perception of visual objects in the game and the gameplay experience, hence the inter-contextuality. These four dimensions are expanded by him in *Table 2.2* (*external vocalization* is divided into in-game voice chat or the use of external Internet radio communications):

Vocalization	*Sound FX*	*Ambient FX*	*Music*
Internal (avatar) External (personal)	Interface (interaction) Object (action/interaction)		Styles: -Thematic -Ambient

Table 2.2. Folmann's categorization of digital game sound.

Internal vocalization is further broken down into:

Speech	*Non-lingual Sounds*	*Singing*
Physical properties: -Age (child, teen, adult, old) -Sex (male, female) -Language: --Dialect / accent --Intonation (emotional expression)	Sighs, Gulps, Laughs, Screams for example	

Table 2.3. The classification of *internal vocalization* in digital games according to Folmann.

Once more, however, with some exceptions as they relate to speech, this is a taxonomy based upon means of production, the source of the sound within the game and the organization of sounds within the game code. It says little about the perception of the sound by the player or the semantics of the sound.

Friberg and Gärdenfors (2004), in an article describing the design and implementation of three audio-only digital games, present the following taxonomy of game sounds as a way to "emphasise the differences between various auditory messages" in their games:

Avatar sounds	*Object sounds*	*Character sounds*	*Ornamental sounds*	*Instructions*
Footsteps, for example, made by the player's avatar. Usually in centre of sound field.	Indicate the presence of objects.	Non-player character sounds.	Ambient music, atmosphere sounds not directly connected to the action.	Such as speech or advice.

Table 2.4. The classification of sounds in audio-only digital games according to Friberg and Gärdenfors.

Such a taxonomy bears similarities to Folmann's in that it mainly describes the organization of groups of sound within the game code. Because of the focus of this taxonomy, there is no indication of possible relationships between sound and image but there is also no mention of the ability of sound to indicate the qualities of either game objects or the virtual space other than a brief mention of 'atmosphere sounds'.

Drawing upon the work of film sound scholars such as Chion and semioticians such as Scott McCloud, Friberg and Gärdenfors derive a triangular model for digital game sound that indicates relationships between a further taxonomy of game sound (categorized initially by the means of production) and modes of listening.[3] For the authors, this model serves as a shorthand rubric for the audio design of their games:

3 The article unfortunately refers to the causal listening mode as the casual listening mode throughout.

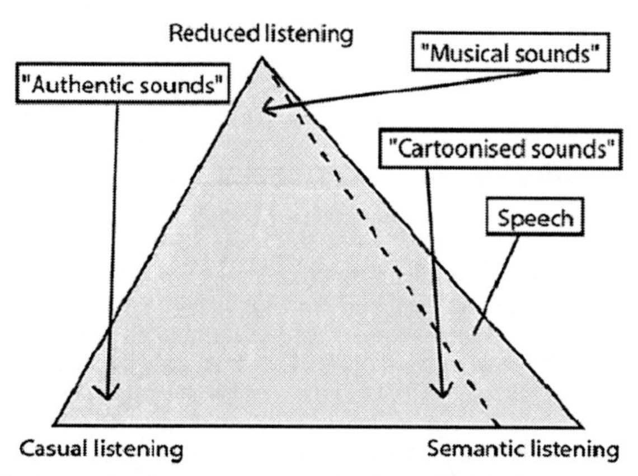

Figure 2.2. Friberg and Gärdenfors' triangular Model of Game Sound (the dotted line is the border between speech and non-speech sound).

This model (which is discussed further in chapter 4) indicates an awareness of modes of listening; that is, that there are different experiences and inferences that may be drawn from sound depending upon the mode in which sound is listened to, and therefore bears some similarity to Stockburger's classification system — for example, the separation of musical sounds from other sounds and the use of the same three modes of listening. *Figure 2.2* is closely modeled on the audio-only games the authors discuss in their article. With this in mind, the important and potentially useful points to note are the clearly demonstrated parallels between modes of listening and types of sound. The mode of listening is not intended to be exclusively related to the type of sound but serves to indicate the first or main mode that will be utilized upon hearing a sound in the game. The implications of this as they relate to the FPS game acoustic ecology are discussed further in section 2.2.

Röber and Masuch (2004), describing their work in designing the audio interfaces of virtual auditory worlds (including audio-only games), take a slightly different approach to Stockburger and Folmann and provide several layers of sound taxonomy. The initial layer, like that of Stockburger and Folmann, is one based upon means of production and game engine processing of sound and divides sound into either story-related sound textures or environmental sound textures (which are the sounds produced when the story-related sound textures are used and processed by the game engine for different parts of the game environment). Categorizing by means of production, they further supply the categories of *speech*, *music* and *natural/artificial* sounds where speech is used to transmit knowledge (tips and tricks on using the game or game mission instructions for example) and music is suited to affect and emotion:

Speech	*Music*	*Natural/Artificial*
Transmit knowledge	Affect and Emotion	Environmental information Object identification -Main sounds -Supporting sounds -Ambient sounds

Table 2.5. Röber and Masuch's functions of sounds in virtual auditory worlds.

The last class, natural and artificial sounds, they further break down into *main sounds*, *supporting sounds* (no further information is given about these two sets) and *ambient sounds*. However, they then reformulate the categorization of natural/artificial sounds into *describing sounds* that provide information about the environment and *artificial sounds* that are used to identify objects in the game.[4]

Leaving aside the obvious question as to why one group of sounds can be classed as having both natural and artificial sounds and can then be reclassified as having both describing sounds and artificial sounds (logically, natural sounds are describing sounds but this connection is not made), Röber and Masuch's taxonomies are interesting in that they go beyond classifying game sound by means of production, by source of the sound within the game or by organization of sound within the game code. Designing sound for audio-only games, they are able to suggest other taxonomic layers such as the information sound can provide to the user or the emotional effect some sounds, such as music, can have on the player.

Breinbjerg (2005) explores the sound of *Half Life 2* (Valve Software, 2004) from an aesthetic viewpoint providing a different approach to the writers above. His article briefly discusses relationships between sound and space and sound and place and touches marginally not only on the semiotics of sound but also on theories of the soundscape as defined by Schafer (1994). It has particular interest for this work because, albeit superficially,[5] he suggests "[a]dopting an ecological approach to auditory perception [in the game as a means to understanding] our natural way of listening" (Breinbjerg, 2005, p.3). Breinbjerg provides three dimensions of space that are constructed by sounds in *Half Life 2* (*Table 2.6*):

Architectural Space	*Relational Space*	*Space as Place*
Defined by qualities of the sound that betray the materials and dimensions of the surrounding architecture.	A dynamic space that specifies the relative positions of the player and the sound source.	The *genius loci* that signifies the type of space or what function it has.

Table 2.6. Breinbjerg's game spaces in *Half Life 2* which may be indicated by sound.

The work returns to these dimensions later in chapter 6 where they are compared to theories of resonating spaces and paraspaces in the FPS game acoustic ecology,

[4] Because the authors are describing audio-only games designed for the blind, they raise the important point that if a game object does not emit sound, it does not exist for the player. In other words, the initial purpose of sound in this type of game, and perhaps its most important purpose, is to reify game objects, to allow the player to perceive them.

[5] In the context of a short article.

but, here, it is worth noting that although these are not taxonomies of digital game sound, they are useful as an initial approach to understanding the role that sound plays in the creation and perception of the variety of spaces that exist within the FPS game's acoustic ecology (this is discussed further in section 2.2).

The taxonomies summarized above provide a starting point for the investigation of the hypothesis. However, in part given the brevity of the articles surveyed, the taxonomies presented are, in the main, cursory and suggest the need for further work. Although there is a recognition that digital games are a different medium, the taxonomies tend to be based upon one strand of theoretical thought (such as film studies) and none of the writings take or substantially develop the multidisciplinary approach that this work, given the space that is available, can afford. The following section, therefore, discusses the writings above firstly as a starting point for the development of a conceptual framework for the study of the FPS game sound and, secondly, from the point of view of their relevance to the hypothesis that the sound of the FPS game is an acoustic ecology in which the player is an integral and contributing component.

2.2 The writings and their relevance to the work

In chapter 1, I proposed a set of basic questions to serve as the basis for selecting and developing existing theories for the derivation of the conceptual framework of FPS game acoustic ecology. The discussion and selection process below proceeds, therefore, on the basis that none of the concepts underlying the taxonomies described in section 2.1 will be applied in the research should they prove irrelevant to an understanding of the FPS game acoustic ecology.

All these taxonomies, with the exception of Breinbjerg's, are based upon either the means of production of the sound (speech or music for example) or, more usually, their organization within the game code (for example character sounds or interface sounds). Stockburger's, Folmann's, Friberg and Gärdenfors' explanations of their taxonomies follow a similar phenomenological or technological theme; that is, they describe either the game code organization of the sound or its means of production but say little about the perception of the sound by the player or its role in the vivification of the game world (or, indeed, how it accomplishes this). Röber and Masuch do impute an affect to some of their classes but this refinement implies, for example, that any sound that is not music is incapable of producing an affect or emotion in the player. Directors, sound designers and audiences of horror films know this not to be the case and many FPS games do include affective, non-musical sound.

Classifications of digital game sound according to their game code organization or the means of production are important to the hypothesis and I therefore develop this taxonomy further in chapter 3. The former classification is potentially a useful source of information as to how the game sound designers themselves view the purpose of the sounds they place in the game. It is often the case (as I demonstrate in chapters 3 and 6), that all sounds whose primary purpose is to provide an indication of the place or temporal period, for example, are stored together within a common set of directories. An examination of this designer-implemented categorization may help explain a variety of intended functions of sound in the FPS game acoustic ecology;

which sounds indicate the presence of other player-characters or game objects or which sounds provide an indication of spatial dimensions, for example.

The means of production as a taxonomic basis (as used, for example, by Stockburger and Folmann) has particular implications for the acoustic ecology creation process. Most sounds in modern FPS games (particularly of the run and gun sub-genre) are recordings of sound[6] and this has several disadvantages compared to using real-time synthesis for the creation of sounds during gameplay (see chapter 3 for further discussion). The main disadvantages are that it limits the number of sounds which may be distributed with the game (and, therefore, which may be heard during gameplay). Furthermore, following on from the first point, the use of recordings precludes the provision of a sound that is unique to the particular gameplay circumstances at any one point in time. The uniqueness of any sound in a real-world acoustic ecology is one of the distinguishing features of such an ecology compared to the acoustic ecology found in FPS games. As an example, consider the tapping of my fingers on this computer keyboard. Each strike (and the perception of that strike) is unique and its sonic characteristics depend on many factors such as the strength of the strike, which finger I use, its temporal relationship to preceding and following strikes or any other simultaneous sounds, the size of the key on the keyboard, its position within the keyboard, the amount of dust or food particles beneath each key, the amount of use each key has had in the past, the material qualities of the table on which the keyboard rests and the material qualities and dimensions of the room in which I sit to name but a few.

Although the more modern FPS game is often capable of providing real-time processing of game sounds reflecting the game environment in which those sounds occur,[7] this is still primitive in the range of sonic difference it provides compared to what is available in a real-world ecology. Ultimately, no digital storage medium is (or is likely to be) capable of meeting the requirements of a real-world acoustic ecology and the solution may well be one that relies on real-time synthesis supported by increases in computer processing power.[8] It is thus important to bear in mind, then, that the mode of production of the sound has implications for the efficacy and the believability of the game's acoustic ecology and that this will impact upon the player's sense of immersion and participation in that acoustic ecology. However, none of the writers discussed in this chapter develop the interplay between mode of production of sound and the viability of the acoustic ecology of the digital game.

Several of the writers discussed above (for example, Stockburger and Friberg and Gärdenfors) make reference to modes of listening (causal, reduced and semantic listening). An important focus of this work is on the player relationship to FPS game sound; the meaning that is derived from the heard sound will affect (and, in some cases, effect) the player's perception of immersion in the acoustic ecology (and, by extension, the game world) and players will have varying responses to sound in the game. Players understand and utilize relationships between themselves and other players and between themselves and the game world in order to partake in the game and a significant channel through which this understanding takes place is through sound. I make the case in chapter 4 that different players, and the same repeat

---

6 That is, they are stored as audio samples on the game distribution medium.

7 This is discussed in chapter 3.

8 I briefly return to this in chapter 5. Additionally, in chapter 7, I question the necessity for emulation rather than simulation.

player with a greater level of familiarity with the game and its sounds, will listen to sound through different modes. Perhaps they take a pleasure in the heard sounds, perhaps they use sound to contextualize themselves within the game world or as feedback on their actions (a panting breathing in some FPS games is a good indicator of the player's energy level) or perhaps they will use it to derive the cause of the sound, the character or the game object that sounds the sound. In the context of the short articles reviewed here, none of the authors fully explore the consequences of utilizing different listening modes for the player's relationship to the game's sounds other than to state briefly that certain listening modes will be prioritized during gameplay. The larger scope of this work allows me to investigate in more depth the relationships between modes of listening and the FPS game acoustic ecology and I do so in chapter 4.

Stockburger, Röber and Masuch and Breinbjerg make note in their taxonomies of the abilities of sound to provide information about the environment and this is certainly important to this work. Other than Breinbjerg (and, to a lesser extent, Folmann), none of the writers explain how sound operates either alone or in conjunction with image in creating the illusion of the 3-dimensionality of the FPS game world. If the player is to have a perception of immersion in that world, it is important that the sounds have a role in drawing that player into the 3-dimensional space of the FPS game world. In terms of image, this is not a volume but, in terms of sound, it is. For a sound to be heard, it must occupy an expanding volume in space and so, in chapter 7, I make the case that the player is physically immersed within the acoustic ecology of the FPS game and that this, in large part, is a factor in the player's perception of immersion within the game world. As a consequence, in part, of this enveloping immersion, the player becomes an integral and physical part of the game's acoustic ecology.

Another factor that contributes to the perception of immersion in the game world, by dint of being able to participate in the acoustic ecology, is the ability for the player to contribute sounds himself. Most of the writers referred to above make note of the player's ability to initiate sounds and I develop these ideas further in chapters 5 and 7. In particular, I note differences between proprioceptive and exteroceptive sounds as created by the player and the effect these have on perceptions of immersion. I further expand this (in a way that none of the above writers do) by exploring the player's experience not only of these self-initiated sounds but of sounds initiated by other players within the multiplayer game world by adapting the concept of diegetic sound (chapter 7).

None of the writings discussed above (excepting Breinbjerg) explicitly discuss the idea that sound can not only provide information about the volumes and materials of the game world but can also provide a sense of location and a temporal meaning (this last is not mentioned by Breinbjerg). By 'sense of location' I mean, for example, the use of sound to convey the setting of the game as London or that the action is taking place within a large cathedral. By 'temporal meaning' I refer to the abilities of sound to convey not only a sense of the progression of time within the game (which may be real time or game time) but also to provide a sense of a particular period of time (past, present, future or more specific periods). Some spatial functions of sound, which I refer to as *choraplastic* and *topoplastic* functions in chapter 6, are hinted at in the above taxonomies with terms such as 'zone object sounds' and 'atmosphere sounds' but are never developed into a taxonomy that fully explores the range of possibilities offered, and none of the taxonomies refer to the temporal functions of sound (which, in chapter 6, I refer to as *chronoplastic* and *aionoplastic* functions of sound).[9]

The player experience of sound is fundamental to the perception of immersion and participation in the FPS game acoustic ecology and several of the writers discussed mention this in passing. Röber and Masuch, for example, suggest that music[10] can bring the experience of affect and emotion to the game but make no mention the abilities of non-musical sound in this area. However, none of them discuss prior experience or knowledge, or cultural and societal conditioning as having an effect upon the player's relationship to sound. Speech classified by its means of production, for example, belongs to an unambiguous set;[11] speech classified as information-bearing sound will be subject to a range of social, cultural, linguistic and other experiential factors (not least of which is an ability to understand the language) with the result that, to some players, the meaning of the speech in the game will be clear (enabling them to engage more quickly with the game) while, to others, it will not be (hampering their effectiveness within the game). Furthermore, the sonic experience, the use that may be made of sound and, therefore, the ability to interact sonically with an acoustic ecology, will not only be different for different players but depends also on the player's familiarity with the game's sound; it takes time to learn the meaning of many sounds in the FPS game. This idea, as it relates to the FPS game's acoustic ecology, I develop further in chapter 4.

Although Friberg and Gärdenfors' triangular model relates modes of listening to sounds classified according to means of production (*Figure 2.2*), which may be of value as a means to designing the audio-only games they discuss, I discard this as being incorrect in many instances if applied to FPS game sound.[12] Their model lacks flexibility in that it does not indicate that the modes of listening that are utilized for various sounds within the game are often dependent upon the player's experience, socio-cultural background and familiarity with the game or that some modes may be used simultaneously. Upon playing a new FPS game, my approach is to first take an enjoyment in the presented sound (reduced listening mode) whilst I perceptually categorize the range of sounds I hear depending upon their sound qualities (reduced listening mode again) and any visual or gameplay events associated with them (causal listening or semantic listening). This multimodal listening approach is used regardless of whether the sound is authentic sound, musical sound, cartoonized sound or speech (as per the authors' classification system). During repeat playings of the game, I come to rely more upon the latter two modes because I am now better able to supply, for example, a cause to the heard sound. Additionally, I suggest that there is a fourth mode of listening which I term *navigational listening*. Sound, particularly in the FPS game, may be used as a means to orientate oneself within the game world and, in some cases, may be used as an audio beacon especially when first learning the game and this is one of the methods by which players are able to

---

[9] All too often, it seems, time is ignored when discussing sound. Perhaps this is because time is taken for granted as sound would simply not exist without it.

[10] Bearing in mind that they are discussing audio-only environments.

[11] Assuming this is recorded speech; synthesized speech, classified by means of production, belongs to a different set.

[12] And, if applied further afield, for example, to music, there are further errors. While it is often the case that music may be listened to and appreciated according to the texture and grain of the instruments (utilizing the reduced listening mode therefore), it is more usually the case (*certainly* the case I would suggest for musicians) that it is also listened to semantically through the agency of musical codes.

contextualize themselves within the acoustic ecology of the game. I discuss modes of listening and the new fourth mode in chapter 4.

Where the writers under consideration make it clear which discipline they base their concepts and taxonomies upon, they utilize and expand upon theories and terminologies derived from film sound and virtual environment design, typically the former. As I made clear in the introduction, film sound theory is a useful starting point for devising digital game sound theory because of a number of similarities. I detail this mainly in chapters 3 and 5 but, briefly, they are both (usually) audiovisual media, are both (usually) popular art forms and both (usually and certainly in the case of popular commercial cinema) provide a stage for human performance (see chapter 9). And there are other similarities. As such, the appropriation of terminology such as the modes of listening, diegetic and nondiegetic sound and acousmatic and visualized sound makes perfect sense.[13] However, there are such significant differences between the two media that an over-reliance on film sound theory leaves many questions unanswered. For example, Stockburger describes the kinaesthetic control players have over the sounding of many sounds and the way they sound (p.9) yet there is no concept of kinaesthetics within film sound theory because (player) interaction has little relevance in the context of cinema. As regards the FPS game acoustic ecology, film sound theory and terminology alone cannot answer questions relating to immersion in the game world, interactivity with the game world, participation in the acoustic ecology and relationships between players and between players and the game engine on the basis of sound, for example. The following chapters explore the usefulness of other disciplines in the construction of the conceptual framework that is key to an understanding of the FPS game acoustic ecology.

2.3 Conclusion

Game Studies is a relatively new research discipline and this is reflected by the limited amount of material and by the types of writing on digital game sound. Those writers who have engaged with the theoretical aspects of digital game sound to date, have provided basic taxonomies of game sound which may be based upon means of sound production and organization or, to a lesser extent, the player's sonic experience, the relationship between sound and image and the spatial qualities of digital game sound. However, such taxonomies generally say little about the role of sound as a component of the game world whether that game world is 2-dimensional as in platform games, or is a 3-dimensional illusion as in FPS run and gun games. Nor do they say much about sonic relationships between players and between players and the game engine. Where such writing exists (Breinbjerg for example), it is brief and inconclusive but, nevertheless, does provide a point of engagement for this research in addition to providing some validation of the *modus operandi* of the work. Additionally, the authors typically utilize the concepts of other disciplines, in particular film sound theory; concepts which include modes of listening, acousmatic sound and visualized sound, for example. All of these are valid points which I develop further in successive chapters but, by themselves, are too sparse and blunt to be capable either of providing a more in-depth analysis of digital game sound or of being the extensive precision tools necessary to investigate the hypothesis. In order

---

[13] Although some of these terms were originally formulated in other disciplines, such as electro-acoustic music, many have been heavily theorized in film sound theory.

to furnish these tools and to more fully investigate the role of both sound and player, research fields other than the few utilized in the writings above must be employed.

The following chapters, then, are the outcome of an extensive research into the literature of a wide range of disciplines and an examination of their relevance to the hypothesis of the FPS game acoustic ecology. In the course of this, I expand upon the taxonomies above and, informed by the fields covered in these latter chapters, devise new and more inclusive concepts of game sound in order to aid in the formulation of the conceptual framework. Although I do further discuss and expand upon all the points of relevance noted above, including matters of a more technical nature such as the means of production of sound or its organization within the game code,[14] I am more concerned with the interpretation of that sound once it is heard by the player for whom it is intended and with analyzing the possible sonic relationships between player and soundscape, player and player and player and game engine. Questions of this nature are not answered fully by the research detailed in this chapter but do form the basis for a multifaceted conceptualization of the FPS game acoustic ecology.

---

14 See chapter 3 which debates such aspects as they relate to the hypothesis.

Chapter 3

Sound Design and Production

Introduction

This chapter looks at issues of sound design and production in the design of digital games where such practices have relevance to the hypothesis and where this relevance is predicated upon understanding the technical possibilities of, and limitations imposed upon, the FPS game acoustic ecology by the game design environment. Before the player purchases the game and enters the game world, experiencing and triggering sounds, these sounds have been chosen and designed by the sound designer. Such choice and design is for a purpose which may be for a sonic indication of a particular game world event or may be for more general or affective purposes. *Pace* Barthes (1967), I make no argument for the authority of the author, the sound designer in this case, but I do make the case that it is worthwhile investigating the practices of sound design and production in FPS games as these may have relevance to the formulation of the conceptual framework. Although I discuss the player's experience and interpretation of FPS game sounds in chapter 4, sound is first designed before it is heard by the player and so such design is discussed here. The player's interpretation and experience of sounds is often based upon sonic conventions (alarm signals are good examples of such conventions and, indeed, are widely used in many FPS games) and, I argue, sound designers make use of these conventions in their work. Furthermore, the use of particular forms of technology for the design and production of FPS game sounds often dictates what is possible and what is not. Is the sound designer constrained to providing a pre-processed sound or is the FPS game engine capable of real-time synthesis or sound processing in a manner that matches, for example, the game world context of the player triggering that sound? Such considerations, and the sound designer's solutions to them, affect the player's experience of the FPS game sounds and so should form part of the conceptual framework.

A large part of the chapter is a comparison between sound design and production practices in the cinema and sound design and production practices in digital games although it by no means surveys the subject to the same breadth and depth of such analyses of film sound production and industry practices covered by books like *Sound Theory Sound Practice* (Altman, 1992). Both being audiovisual media, comparisons are often made between the two industries (see chapters 2 and 5) and there is often crossover and migration of creative personnel from one to the other as I explain below. It seems unavoidable in these cases that some of the practices of one industry will be co-opted or syncretized with another. As in popular commercial film, sound design and musical compositions for digital games are often almost afterthoughts and are bolted on at the post-production stage (see Thom, 1999; Bridgett, 2003; Guy, 2003 for example). Even in the rare cases where sound design is considered from the inception, the sound designer usually works to many constraints dictated by the game coders. In response to a question about his status, Mark Klem, the sound designer for *Urban Terror* (Silicon Ice, 2005), gives this response:

> If I remember correctly, I came in when things started to really begin. There was already a small base of the mod, but it was really only a few weapon tweaks and things that BotKiller had done. After I got on and we got some new coders and modellers, things started to really change. So the end product was actually put together by everyone at the same time [...] We always wanted more actors for different characters, however, the coders let us know very quickly that memory was the real issue here (Klem, personal communication, 2004).

The aims of this chapter, then, are to briefly describe the sound design processes that are a part of the FPS game design process, to make comparisons between cinematic and digital game sound production practice and to provide some preparatory work for more focussed research in later chapters. In particular, this is the case for chapter 5 which takes the relationship between sound and image and sound and event in the cinema for its starting point in debating these relationships as they pertain to digital games. Although the literature surveyed in chapter 2 in some cases makes use of theories borrowed from film sound theory (Friberg & Gärdenfors, 2004; Stockburger, 2003), none of the authors discuss commonalities and differences between sound design and production practices in the two industries.

Importantly, though, a section in this chapter also looks at the organization of audio samples on the FPS game distribution medium. For this work, it is important not only to analyze the role of sound during the playing of the game and how it relates to the player experience but it is also, I suggest, important to look at the way in which sound is classified by the FPS game sound designer. At the very least this provides an initial taxonomy for the conceptual framework that is being devised throughout this work, but it also provides an insight into the intentions of the game sound designer. It is assumed that the purpose of sound design is to provide a particular meaning for the player in the game and that, initially and at a fundamental level, this purpose may, in part, be illuminated by such an analysis. Accordingly, I investigate the naming and distribution of sound directories and audio samples using *Quake III Arena* (id Software, 1999) as an example and as a preparatory exercise for the work in chapters 8 and 9 which uses *Urban Terror*.[1] Again, none of the authors discussed in chapter 2 fully analyze the organization of audio samples on the distribution media[2] of the specific games they describe.

This is also a chapter which, albeit in a cursory manner, discusses some aspects of the digital game musical soundtrack. Although this work is about diegetic sound in the FPS game, discussions of the musical score are a useful entry point for raising similarities and differences in sound practice between the design of sound for digital games and the design of sound for films.[3] A useful historical survey of the technology of sound reproduction in a variety of gaming consoles and desktop computers can be found in Collin's work in progress (Collins, n.d.) — this chapter,

---

[1] *Urban Terror* is a total modification (mod) of the *Quake III Arena* game engine but, as a 'realism mod', it proposes a different scenario to the more fantastical *Quake III Arena*. The use of two diverse examples throughout the thesis helps to broaden the application of the conceptual framework and hypothesis.

[2] The compact disc, DVD-ROM or hard drive on which the game is installed.

[3] Separating them into 'two industries' is an arguable point given recent convergence between the two exemplified by business acquisitions and mergers which group representative companies from the two industries together in the one conglomerate; even more so when games are reduced to film marketing material or vice versa.

given the nature of the work, concentrates on current technology that is used in the design and production of sound for FPS games and on the organization of sound within the FPS game.

3.1 Hardware and software and comparisons between the digital games industry and the film industry

This section briefly discusses the hardware and software that is used in the design and production of sound in the FPS game and makes some comparisons between sound design practice in the digital games industry and in the film industry. All of these design stage factors have a potential effect upon the FPS game acoustic ecology as experienced by the player and, as is outlined in chapter 2, none of them have been discussed by other digital game sound theorists. Where descriptions of game sound design hardware and software may be found, they concentrate solely on the production practices of game sound design and do not relate these to the player experience (Brandon, 2005; http://www.gamasutra.com/, for example).

Firstly (and comparing the film and digital game industries), a large amount of equipment used (both hardware and software) is either the same or similar in input, output and intent. Thus there are recording and mixing consoles, MIDI and audio sequencers, microphones, FX processors, compressors and gates to name merely a small fraction of the technology to hand. Furthermore, much of the terminology of film and game sound practice is the same. Hence we have, for example, dubbing, voice-over, soundtrack, sound FX and sampling and this points to some similarities not only in the tools used but also in the way those tools are used between the two industries.

Secondly, some of the sound personnel involved (whether freelance or studio personnel) do similar work in both film and games. Although this work is not focussed on nondiegetic music in FPS games, this is best illustrated by looking at the way music composers freely shift between film composition and game composition. Michael Giacchino is a composer who originally worked in film, then worked on games, including FPS games such as *Medal of Honor* (1999-2006), before more recently switching back to film music including composing the music score for *The Incredibles* (Bird, 2004). Jesper Kyd is a composer for the *Hitman* (2000-2006) series of games and more recently for the horror film *Stranger* (Hess, 2005). This is not uncommon and, at first sight, this adaptability and flexibility seems admirable, but there is a whole class of computer games (usually of the role-playing epic type such as *Neverwinter Nights* (Bioware, 2002)) for which Cavalcanti's (1939) scathing comments (written during the early talkie era) about the use of neo-romantic and neo-classical pastiche in cinema could easily be applied to:

> [I]t is an idiom suited to an atmosphere of pomp and display. In style, the music of the cinema, by and large, represents a fixation at a stage of development which the art itself left behind about thirty years ago. Tschaikovsky, Rachmaninov, Sibelius, are the spiritual fathers of most cinema music.[4]

---

[4] And it continues to this day. Perhaps nothing better demonstrates the Hollywoodization of Peter Jackson than the musical scores for the *Lord of the Rings* trilogy (2001-2003) and his remake of *King Kong* (2005) which, by comparison with the music utilized for his idiosyncratic early horror films, represent the apex of Hollywood's continuing

And this from David John, a composer involved in *Neverwinter Nights* in response to the interviewer's request *[w]ithout thinking, name 5 bands/composers that everybody should listen to*: "Stravinsky, Shostakovich, Copland, Dvorak, and of course Holst. So much great film and game music is influenced by these composers, especially Holsts' [*sic*] The Planets" (Watamaniuk, n.d.). Readers familiar with Western art music will recognize stylistic and temporal similarities between the two lists.[5] However, the focus of this work is not musical pastiche, although it would make for a potentially interesting study on the oddities of hearing late nineteenth and early twentieth-century Western symphonic music, as mediated by popular commercial film practice, in a range of digital games.

Finally, and this is related to both points above illustrating commonalities of technical tools and their usage and a similar use of sound design technique (at least in the initial gathering of sound), sounds themselves can cross over from one medium to the other. Mark Klem, the sound designer for *Urban Terror*, says that most of the weapon sounds "came out of Bryan Watkin's [*sic*] personal library in his Hollywood Studio. He works for Warner Bros." (Klem, personal communication, 2004). Of interest here, is that among Bryan Watkins' many credits as sound designer or sound editor is one for the recent film *Doom III* (Bartkowiak, 2005) based on the FPS game series of the same name.

A difference in sound design practice between film sound and FPS game sound is found in the manner of recording and reproduction of sound. Film sound (and here I ignore film sound prior to the advent of the talkies and attempts to co-opt other external factors such as audience noise into the canon of film sound) consists almost exclusively of location-recorded sound (which may be synchronized to image later) or dubbed sound.[6] This latter sound may derive from a recording of a suitable sound for the action (specifically recorded for the film or from a library of pre-recorded sounds) or may be created (that is, synthesized). Sound may be used in a verisimilitudinous manner or not and a sound recording, whether it is used to suspend or engage belief, may be taken from the sound source as viewed on screen or may derive from another source entirely unconnected with the screen source. FPS game sound (and all digital games have similar technological options or limitations in the derivation of their sound entities), has no location recording in the sense of recording sound on the set or on the location as filming occurs. It is therefore possible to state that location recording is (currently) impossible in digital games and this has both practical and semantic implications as will be discussed below.

Unlike film sound practice, all digital games (including FPS run and gun games, although it is little used if at all) have the technological option to synthesize sound in real time. Furthermore (and this option *is* widely used in the more recent FPS games such as *Half Life 2* (Valve Software, 2004)), sounds can be processed in real time depending on the environmental geometry and virtual materials of the player's

---

infatuation with the romantic and neo-classical composers. (The fact that Richard Wagner's music is used uncredited in the final installment of the trilogy and that it represents little stylistic change to Howard Shore's music is a good evidence for this.)

5 FPS games are not immune to this charge — *Battlefield 1942* (Digital Illusions, 2002) has introductory music that is a poor imitation of Holt's *The Planets* Suite (the *Mars* movement of course). There is also a relationship between digital games and the popular music industry and, in the case of FPS games, this is exemplified by the use of compositions by Trent Reznor (of Nine Inch Nails) for *Doom* (id Software, 1993).

6 Allowing for the fact that recorded sound FX may be so heavily processed as to sound synthetic or that synthesized sound FX, although rare outside of genres such as science fiction cinema, do occur.

location within the game. In other words, and as an example, the player's footsteps on a similar surface (perhaps concrete) will have different reverberation and other acoustic properties depending on whether the player is within a train station, in a small room or is walking about outside even though the basic sound that is used (digitally recorded or created footsteps on concrete) is the same in each case. Sound is further dynamicized in FPS games by a degree of responsiveness and flexibility to players' actions which means that the sound heard in the game will not only be different for each player at the same point in time but will also be different for any one player at each playing of the game. A large part of this is to do with the player-character's physical (and virtual) relationship to the sound in the game — the relative positions of the sound and the player's character affects the directional properties of the sound and its perceived distance — but it also refers to the fact that the sounding of FPS game sounds in large part depend upon player actions which will be different every time the game is played.[7]

This real-time processing of the FPS game's audio samples (and the potential for real-time synthesis) is a start in overcoming the limitations in the FPS game's acoustic ecology. Without such real-time processing, the palette of sounds available is limited by the storage space given over to audio samples on the game's distribution medium. Although *Quake III Arena* has almost 600 audio samples (as noted in section 3.2) this pales in comparison to the almost infinite range of sounds to be found in natural ecologies. This is not to say that natural ecologies are at all times replete with an infinite variety of sounds but that any particular sound in that ecology (a footstep, for example) will always sound different each time it sounds depending upon not only changes in the way it is sounded (in the case of the footstep, a heavier step, a lighter step, a stumbling step for example) but also upon the basic interaction of sound with a range of environmental factors (such as the material the foot is walking upon, the reflective properties of the surrounding objects, the volume of any enclosing space). Furthermore, it is highly likely that such storage limitations (particularly in the older games distributed on less high-capacity media) also limit the types of sound in the FPS game's acoustic ecology. Not all FPS run and gun games include speech (*Doom*, for example, had no speech) but many of the more modern ones do (such as *Battlefield 1942*). This is probably due to the increased storage capacities of FPS game storage media (audio samples of speech typically require more storage space than short sound FX samples).[8] The sound designer for *Urban Terror* has this to say:

> I know we spoke many times early in development about the possibilities of having a few totally different sets of voices to use for different characters. Many with foreign accents. This was never going to happen due to the memory constraints and large download size (Klem, personal communication, 2004).[9]

---

[7] This is discussed further in chapters 7 and 8 and, for example, may be to do with the player's choice and use of weapons being different at each playing of the game or different gameplay patterns as a result of different opponents.

[8] There may also be aesthetic reasons for this too. Arguably, single-player games have less need for communicative speech than do multiplayer games.

[9] *Urban Terror*, being free software, is typically obtained through an Internet download (v3.7 is 446.6MB) hence the concern over file sizes. Games distributed on CD-ROMs may store approximately 700MB and games distributed on 8cm DVD-ROMs can use between 1.4GB and 5.2GB depending on the DVD format. Such continued increases in storage capacity can only have beneficial implications for the FPS game acoustic ecology as will be clear from comparing the primitive sound usage of early digital games to the multichannel sound use of modern games. Chapter 7 does, however, question the need for emulation of natural acoustic ecologies.

Such real-time synthesis and digital signal processing (DSP) is not available to film viewers once the film is released (naturally, real-time DSP is available during editing and post-production and real-time DSP in the area of noise reduction and surround sound processing is a feature of many cinema playback systems but these lack the ability to process individual sounds). Once the release print has been made, the recorded sound on that print is fixed and immutable. The final film soundtrack is one entity, one totality and in this differs substantially to digital game sounds where the soundtrack is created during gameplay and is created not only differently for each player but also differently at each playing.[10] (The same may be said for the screening of a film compared to the playing of a game — one of the reasons why I do not favour the use of narrative as a way of explaining digital games. I discuss this further in chapter 9.)

The sound that is available for playback in FPS games consists of discrete packages of audio rather than one self-contained and unchanging soundtrack as is the case in cinema. FPS game audio may consist of hundreds or thousands of individual digital sound files or audio samples. In this, the practice is not too dissimilar to that used in (diegetic) audio post-production in film where the dubbing process may be entirely digital. Such post-production facilities possess or have access to vast libraries of audio samples which, as noted above by Klem, are sometimes used for purposes other than film. However, the similarity stops here for whereas in film, sound, like image, is created and then stored fully-formed in the final print, FPS game audio samples remain in a pre-natal primitive state until the player's in-game labours give birth to a unique and fleeting soundtrack (see the discussion on sonification in chapter 4).

In-game actions, whether they be the result of player decisions or other factors such as timing events, will trigger parts of the game engine code which in turn load the appropriate audio sample from the game audio library stored on hard disk or removable media (usually the former for speed of access). This audio sample will then be played back through the player's audio system, with appropriate DSP depending on the game engine's capabilities.[11] In most cases, then, the sounds heard in the acoustic ecology are a result of an action whether that action be player input, a bot action, or game status event. The fact that sound in the FPS game is predicated upon action is directly responsible for the use of discrete audio samples as the sound mechanism rather than, as in cinema, a pre-composed soundscape and it also, as demonstrated in chapter 7, has important implications for the player experience of, and engagement with the acoustic ecology, as they relate to interactivity and immersion.

There is no master soundtrack in FPS games. Film sound designers can work holistically and linearly, aware of how each sound item they record or construct and

---

[10] Live accompaniment to silent film may differ from screening to screening depending upon a number of factors (which score is being used, how much improvisation is involved, the orchestration or arrangment or the number and musical style of the performers for example). Yet, at each screening, this accompaniment (if none of the factors listed above has significantly altered) is still substantially the same for each audience and for each member of any one audience — they will all hear the same order of notes, the same harmonies, the same rhythm and dynamics as the next person. In other words, they will hear the same music.

[11] This description is somewhat simplified as it does not take into account other factors; pre-caching of sound, multichannel audio or head-related transfer functions for example, descriptions of which may be found in a range of articles on the *Gamasutra* web site (such as Tsingos, Gallo, & Drettakis, 2003) or in a variety of books (like Dodge & Jerse, 1985; Howard & Angus, 1996; Turcan & Wasson, 2003).

weave into the soundtrack functions not only horizontally in that soundtrack, as part of a sequence with other sound items, but also vertically, layered in the same time slice with other sound items. The FPS game sound designer (with the possible exception of matching, for example, parish church bells to a repeating audio sample of countryside sounds) is forced to work atomistically with the knowledge that the final conception of the game soundtrack is out of his hands. The pool of available audio samples and the provision of background, scene-setting sounds for example, may be decided upon by the sound designer and be under the control of the game engine, but, usually, the playing of these sounds and other character-related sounds is dependent upon player actions during the game. Some background sounds are only played when the player is within a certain location of the game level and other sounds, such as the character's gunshots or footsteps, have to be triggered overtly by the player before they sound.[12]

The process of film cutting and other editing practices is predicated upon the image (for example, early analogue editing systems had no sound capability). According to Doane (1980), and disregarding a significant body of cartoons and music video, in popular commercial cinema, sound is subordinate to, and added to, image because, in cinema, the empirical (the image in this case) takes priority (pp.47—49). FPS games (as far as the actual gameplay within a level is concerned and disregarding spectator modes, overhead views upon in-game death and full motion video (FMV) inserts for example) have no concept of a shot or even of scenes and so FPS sound is freed from this particular dependence on the image and from the subordinate role of unifying a series of images. As will be demonstrated in chapter 5, sound in the FPS game is predicated, in the main, not upon image but upon players' actions.

There are a variety of approaches to classifying FPS game sound by production and one possible approach is based on the means of production of the audio sample. An audio sample may be either a recording of a real-world sound (possibly with further processing by means of DSP) or it may be a synthetic sound. In the first case, such audio samples within the FPS gameworld may be recordings of weapons fire, footsteps on a variety of surfaces, water sounds, wind sounds or speech — in fact anything that the game sound designer chooses to record from the real world to fit the particular FPS game or level. Although I discuss this further in chapter 4, an example would be the *Urban Terror* gunshot samples that are recordings of the gunfire of the actual weapons represented in the game. Such recordings may be digitally processed and it is likely that all audio samples will undergo some form of processing at least to normalize the amplitudes,[13] to resample them to the game audio engine's required bitrate and sampling frequency, to convert stereo recordings to mono samples or to trim the length of the recordings by removing silence from the beginning and end. Other types of optional processing may include compression, equalization or reverberation, for example. Synthesized audio samples,[14] rather than being recordings of real-world sounds, may be created with tools which implement some form of audio synthesis such as additive synthesis, frequency modulation or granular synthesis. Like the sample recordings, they may also undergo a range of

---

[12] There are some global sounds under the control of the game engine, such as sounds indicating a game level has been won or lost, or a flag has been captured, and that are dependent upon the general gameplay resulting from the sum of all players' actions. These will be played and sounded for all participants in the game.

[13] That is, to ensure that all audio samples have the same peak amplitude if they are intended to have equivalent volume.

[14] In *Urban Terror*, fewer in number than recorded audio samples.

processing following the initial synthesis.

Real-world samples tend to be aligned with their virtual world analogue in the game. This is particularly the case with FPS games which attempt some simulation of reality by including analogues of real-world objects — games such as *Counter-strike* (Valve Software, 1999) and *Urban Terror* for instance. Thus, a recording of a SPAS shotgun (or one of several similar audio samples provided in the game for sonic variety) will be played each time the SPAS is fired in *Urban Terror*, samples of footsteps on gravel will be played whenever the character moves about on virtual gravel, samples of birds and wind rustling through the leaves may be played whenever the character is in a forest and samples of machinery may be played when the character is inside a power plant.

Sometimes, this causal, quasi-indexical link[15] is broken when it is found that a particular audio sample is a more appropriate sound for an object within the game that is other than a representation of the real-world object from which the recording was originally taken. This is the case for the more fantastical FPS game's objects and characters for which no real-world equivalent exists — a variety of characters in *Quake III Arena* for example. In this case, the resultant audio sample may derive from several recordings (or be synthesized) and is likely to be so heavily processed that the original recordings are unrecognizable.

Synthesized samples may be utilized in the manner described above or may be designed to be heard above a mêlée of other game sounds because they signify some significant gameplay event. *Quake III Arena* has several of these sounds — indicators that a power-up (such as quad damage) has been picked up by a player for example — and often such sounds are global sounds being heard by all players.

Sound practice in the design of FPS games, therefore, affects the resulting heard sound both through the types of sound that are used and in the use and disposition of sounds within the game. The perception of immersion within an FPS game acoustic ecology requires participation from players and this, in large part, is achieved through individual audio samples sounded in response to player input. Furthermore, this responsive sounding of audio samples on the part of the game engine means that not only will the acoustic ecology be continuously adapting to game events and player actions (autopoietically, the ecology compensates for perturbations in the system), but it also means that the acoustic ecology is different at each playing of the game.

3.2 Game code organization

It is also possible to analyze FPS game sound based on the sounds' organization within the game engine; an analysis that demonstrates, in part, the game designer's method of sound classification. This provides an insight into the intent of the sound designer when designing particular sounds — in other words, what is the purpose of the sound as designed by the sound designer? — and serves as a preliminary taxonomy of FPS game sound in the conceptual framework being constructed

---

[15] See chapter 4 for a discussion of these terms. Briefly, causal sound refers to a sound from which a listener may, with a variable degree of success, divine the origin of the sound and quasi-indexical is a term I use to refer to the virtual indexicality that some digital game sounds possess with respect to in-game events and visual objects.

throughout the work. Here, I will use *Quake III Arena* as a counterpoint to the use of such a taxonomy as illustrated by the use of *Urban Terror* in chapter 8.

Once *Quake III Arena* is installed, it is possible to peruse and play all sounds used in the game and to see their organization into directories on the hard drive. All sounds are stored in the sound/ directory which itself has several subdirectories named for the game designers' sound classification system (there is a separate music/ directory for all the game's musical soundtracks as opposed to sound FX). Thus, there is a teamplay/ directory containing audio samples to do with team-based game configurations, such as speech samples indicating whether the red or blue team has captured a flag. The feedback/ directory contains mainly speech samples indicating, for example, how much time is left in a timed game, samples praising the player who has managed to 'frag'[16] several opponents in a short specified time span, and, a particular favourite, the derisive *Humiliation!* sample that is played whenever a player's character has been splattered by the gauntlet, a close-quarters and difficult to wield weapon.[17] The items/ directory contains samples relating to specific interactable objects within the game that the player can pick up or use such as armour or quad damage. The misc/ directory has samples to do with the menu interface (these samples are not a part of gameplay and so are nondiegetic sounds) and the movers/ directory contains just three audio samples for the sounds of doors opening and closing and switches being operated. The world/ directory contains environment audio samples, many of which are designed to be looped, and that indicate specific locations or levels in the game or that are attached to non-interactable game objects such as fires or fans.

The two largest sound/ subdirectories are weapons/ and players/.[18] In the former, there are general audio samples used when a player drops a weapon or is out of ammunition and further subdirectories within the weapons/ directory for each of the game's weapons, eight in all. Most of these are firing sounds and although some weapons, such as the machinegun, have multiple firing audio samples for sonic variety, others, like the shotgun, have just the one. The players/ directory contains twenty-three subdirectories for each of the characters a player may choose to play as. These directories consist of sounds (as with some of the weapon audio samples, often multiple samples associated with one action are provided for variety) emitted by the character upon dying, falling or taunting an opponent, for example. The top level of the players/ directory has several general samples used equally by each character (such as the sounds of splashing if the character is in the water), there is a footsteps/ directory with samples of footsteps for classes of characters on a variety of surfaces and an announce/ directory consisting of speech samples indicating which character has won a tournament or deathmatch game (only used in single-player configuration, this directory also contains the optimistic sample *You win!*).

This game designer-constructed organization of sound is illuminating in several respects. Firstly, it is an indication of how the game engine deals with sound and its relationship to a variety of characters, objects and locations within the game. Sounds that player or bot characters create as they move, die or taunt are separated from environment sounds that are part of a location, sounds of interactable objects (such

---

[16] In FPS games, to frag is to kill an opponent.

[17] This sample is a favourite among *Quake III Arena* players so much so that the game nickname *Hugh Miliation* is a popular one with gauntlet aficionados.

[18] The existence and size of the former, perhaps, being a good indication of the game's genre.

as picking up health or firing a weapon) are separated from the sounds non-interactable objects (such as fans) make and diegetic sounds are separated from nondiegetic sounds (the latter are in the misc/ folder).

Secondly, the sheer number of sounds (575 in total in the original *Quake III Arena* release) is an indication of the importance of sound to the game experience. Gone are the two or three blips and bleeps of the arcade games of yesteryear; every character and every weapon has its own complement of sounds which work together with environment sounds and interactable sounds to create an acoustic ecology approaching, in number and diversity of sounds, a natural and busy acoustic ecology.[19]

Thirdly, this organization of sound indicates some of the technical limitations of the game namely in the areas of media storage and computer memory. At first sight, it seems curious that there is a separate directory for the footsteps of game characters. It would seem more logical to place unique footstep samples in each of the character's subdirectories. However, many of the twenty-three characters share similar characteristics; Sarge and Bitterman are human characters as opposed to the more robot-like and metallic Tank or the eyball on two legs that is Orb. Some footsteps are therefore shared between classes of characters (for instance the human characters) and this decreases the number of sounds that must be stored on the game's distribution media (a compact disc in this case) and that must be loaded in the computer's memory while playing.[20]

A preliminary taxonomy of FPS acoustic ecology sounds based on the above assessment of *Quake III Arena* may now be proposed. It consists of character sounds, interactable sounds, environment sounds, feedback sounds and nondiegetic miscellaneous sounds. This is a simplification of the directory structure detailed above in an attempt to broaden the application of such a taxonomy to other FPS run and gun games. For example, I group weapon sounds with other object and event sounds such as items and doors and I provide no separate category for speech audio samples preferring instead to place such samples in any of the five categories according to their function. This is because, the hypothesis is best served by a taxonomy that classifies according to function in the game rather than mode of production or the original source for the sound. Nevertheless, it is a taxonomy that does reflect the game code organization and, therefore, the game sound designer's intentions because such intentions (as this examination has shown) are based upon in-game function.

There are, then, some significant points of difference to the taxonomies outlined in chapter 2. The first is an absence of speech as a separate category as is found in the taxonomies of Stockburger, Folmann (n.d.), Friberg and Gärdenfors (2004) and Röber and Masuch (2004). Freeing speech to serve duty in categories such as character sounds or feedback sounds is more indicative of the functions of such speech and therefore the player's engagement with it. It may, for example, be used

---

[19] With the proviso, as noted above and particularly in the case of *Quake III Arena* which has no real-time processing of audio samples, that each audio sample always sounds the same rather than being affected each time it sounds by a range of parameters as would be a sound in a natural ecology.

[20] *Quake III Arena* does allow the user to display all character models (and their sounds) using the same model the player is using in an effort to conserve RAM. However this is not the default setting and, from personal experience, is rarely used.

by a player to taunt another or it may be used by the game engine to provide instructions to players or for feedback on significant game events. Secondly, having a class of interactable sounds (something not found in any of the other taxonomies) allows for audio samples of weapons and other objects, the sounds of which may be triggered by players, to be classed together. This serves as an indication of the level of player-sound interaction in the game — the level of participation in the FPS game acoustic ecology the player is capable of — and this has implications for player immersion within the ecology. It is likely to be the case that the more interactable sounds an FPS game has, the more immersive it is and the greater the participatory potential there is for the creation of the acoustic ecology (chapters 5 and 7). Character sounds and environment sounds serve to distinguish sounds that help player identification with a character from sounds that aurally reify the FPS game world and these help prepare the way for discussions on the acoustic spaces of the game (chapter 6) and player immersion (chapter 7). Stockburger's and Folmann's taxonomies, for example, while having categories of environment sound (or similar), have no separate class of character sounds and, indeed, their taxonomies provide no indication that there are characters (and therefore players) involved in the game at all. For the purposes of the requirement of the hypothesis that the player be an integral and contributing component of the acoustic ecology, a taxonomy of FPS game sound is needed which not only classifies sounds *per se* but also provides the conceptual hooks for investigating the player relationship to these sounds. This, then, is what the taxonomy formulated here intends to be and it forms the basis for the exploration of the hypothesis in subsequent chapters.

3.3 Conclusion

There are several similarities between sound design and production in the cinema and sound design and production in digital games but there are also many differences between the two. The latter are accounted for by the interactive nature of digital games and, especially in the case of FPS games, by the free-wheeling nature of gameplay where the exegesis[21] of the gameplay, between the start of a level and the end of a level, is in large part dictated by player input. Because of this, sound must be predicated, at the design stage, upon game action rather than upon image (as is the case with cinema) — this is discussed further in chapter 5 — and there can be no fixed diegetic soundtrack in digital games. Rather, sounds are stored as discrete audio samples to be used as and when required during the playing of the game. Such practices have significance for the concept of an FPS game acoustic ecology because they mean that the player himself contributes sound to the ecology through his actions and, furthermore, it means that sonically and (assuming player response follows heard sound) ludically, no two playings of the same FPS game will be alike.[22] Furthermore, an acoustic ecology contains no visual images and so consists not only of sound that refers to spaces and materials within the game world (chapter 6), but, more importantly, consists mainly of sounds that refer to player actions and game engine actions. In the case of player actions, such sounds are a valuable indicator of the viability of the acoustic ecology because they are a direct representation of player engagement with that ecology.

---

[21] The plot as opposed to the story.

[22] An FPS game can be played again and again without loss of enjoyment (certainly for FPS aficionados) — repeat viewings (where not only the sound but also the image remain constant) of all but the very best films pall after a while.

The organization of audio samples on the distribution medium is a classification of FPS sound and an analysis of this provides insight into the game designers' (and sound designers') approach to sound. Following on from this, this classification helps to indicate which sounds are attached to which spaces, objects and events within the game world and so helps to tease out some of the basic building blocks of the FPS game acoustic ecology. Based on an analysis of the sounds in *Quake III Arena*, this preliminary taxonomy of sounds in the FPS game world comprises character sounds, interactable sounds, environment sounds, feedback sounds and nondiegetic miscellaneous sounds. Accordingly, the basic model of the FPS game acoustic ecology introduced in chapter 1 may now be added to by inserting this taxonomy as an element of the soundscape.

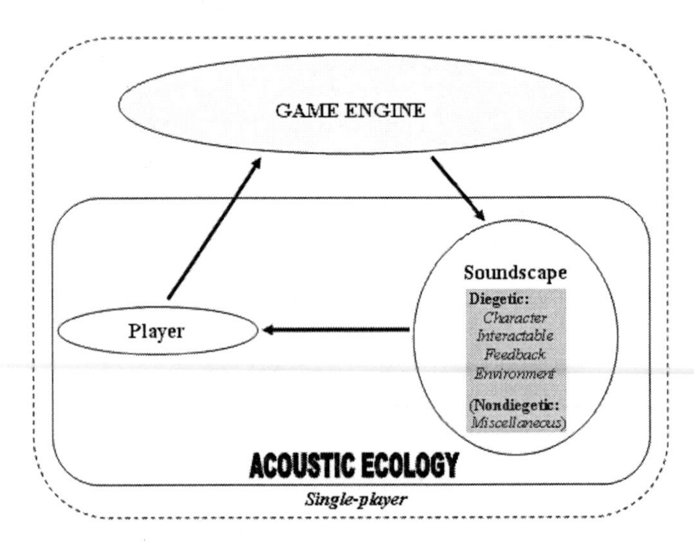

Figure 3.1. A taxonomy of FPS audio samples added to the acoustic ecology model (single player).

As a taxonomy, although it illuminates the organization of sound within the code and goes some way to explaining the game structure and the game designers' intentions, it says little about the function and meaning of sound, how sound is used in the game by the player or how sound constructs the game's acoustic ecology. However, it does indicate some of the basic building blocks from which the acoustic ecology is constructed by providing classifications of sounds within the ecology's soundscape. Although the sound designer's classification of sound ascribes some basic meaning to the sounds, it is important that the player is able to divine such meaning while playing the game. A character's footstep, designed as such by the sound designer, must be interpreted as such by the player rather than, for example, the sound of gunfire. In most cases, the meaning of sound is dependent upon several factors, not least of which is context, but the properties of source object and event are no less important in arriving at this final understanding of the meaning of the sound. Sounds sound differently and have different meaning because their source objects and events are different. A recognition of the basic parameters of a sound, such as pitch, loudness, vibrato, for example, combined with experience in the hearing of that or similar sounds, potentially leads to an identification of the source object and event. How sound, at a fundamental level, is able to provide an understanding of the objects and events it is sourced from is the subject of the next chapter.

Chapter 4

Meaning in Sound

Introduction

Chapter 3 dealt with the practicalities of digital game sound production and the organization of audio samples on the distribution medium. In part based upon the classification that sound designers themselves give sound, a preliminary taxonomy of diegetic sounds in the FPS game was devised comprising character sounds, interactable sounds, feedback sounds and environment sounds. The basic model of the FPS game acoustic ecology was expanded to include this classification in order to illustrate the range of sounds found in the soundscape. Founded as it is on production techniques and the organization of audio samples, the taxonomy demonstrates little about how meaning is derived by the player from each sound in the game or how meaning is designed into sound. This, then, is what this chapter makes a start on — investigating meaning in sound *sui generis*, that is, meaning prior to its modification as a result of the context in which the sound sounds.[1]

The FPS game world is, during gameplay, the player's environment and, while not ignoring other channels of information, it is my contention that, particularly in this type of digital game, it is primarily through sound that the player derives gameplay meaning, is able to contextualize himself within that environment and is able to make his presence known, thereby becoming an integral and contributing component of the game's acoustic ecology. Prior experience and knowledge are important here; for example, an understanding of sonic conventions in a particular culture, such as alarm signals. The ability to map real-world sonic experiences and their meanings on to similar sounds in the game environment and previous exposure to the sounds of the specific FPS game in question or to the wider conventions of FPS game sound all play their part in making sense of this virtual space.

An initial section of this chapter is used to point out the importance of player experience in assessing the meaning of FPS game sound. This experience may be based upon previous socio-cultural factors or may be a result of experience in the conventions of FPS games or the playing of the specific FPS in question. This is a recurring theme throughout this work because this experience affects not only the meaning ascribed to sound at a higher cognitive level but, at a more fundamental level, in many cases is likely to affect modes of listening (see section 4.2) and sonic elements of the acoustic ecology influencing immersion which may depend for their efficacy upon player familiarity with the sound (see chapter 7 for a discussion of immersion). Furthermore, it is important to note the player's use of sound in any particular FPS game is likely to change due to increasing familiarity with, and exposure to, those sounds. This may be described as a reprioritization of the hierarchy of affordances the sounds offer (see section 4.3) and indicates that there is both a conscious and a subconscious training involved as the player actively learns

---

[1] It is, of course, impossible to ignore context when listening or to play a sound in a context-less void. Although I leave aside references to specific contexts (which team member has triggered a sound, for example), here, the reader may assume that the FPS game is the general context in which the sound sounds.

the game and absorbs experiences from it.[2]

It is important to note that FPS game players do not (necessarily) consciously register the fundamental parameters of the sound that are heard. In other words, they do not, upon hearing a sound, make note of its frequency, its duration and its intensity in anything other than the most general of terms (a long, loud, bright and high-pitched sound, perhaps, rather than a 10.6 second duration sound with a fundamental frequency of 3567Hz, a noticeable proliferation of overtones and an intensity of 105dBA). Instead, as in the real world, what is consciously noted may be the cause of the sound, the materials of the sound source or the location of the sound. It is from these parameters that the derivation of meaning proceeds.

Therefore, while not ignoring sound as sensation, the work concentrates on the perception and cognition of sound in the FPS game and this particular chapter concentrates on how players derive meaning from sound. Where meaning can be derived from sound this allows for immersion in the FPS game acoustic ecology and participation and contextualization within it. To investigate this ability within the FPS game genre, I investigate sonification and auditory icon design (which are both concerned with designing meaning into sound and deriving meaning from it) in addition to (and to a lesser extent) other concepts (such as modes of listening and theories of affordance) which prove to be of preparatory use at the start of the chapter. It is noticible that auditory icon design, as it applies to FPS games, is closely related to the mode of production outlined in chapter 3. That is, the method of sound creation or production has importance for the design of auditory icons.

4.1 Player experience

A number of authors from a variety of disciplines note the role experience (and, therefore, the acquisition of knowledge through experience) plays when assessing the meaning of sound (Jones & Yee, 1993; van Leeuwen, 1999; Kress & van Leeuwen, 2001; McAdams, 1993; Walker, 1987 for example). Although this experience is acknowledged by authors discussing player engagement with digital game environments (Kattenbelt & Raessens, 2003; Roudavski & Penz, 2003 for instance), it is not explicitly acknowledged by the authors surveyed in chapter 2. This section, therefore, serves to raise awareness of the importance of sonic experience in the player's engagement with the FPS game acoustic ecology. As a preliminary exercise, such experience may be classed as socio-cultural experience (that is, all sonic experiences not to do with FPS games), experience gained through the playing of FPS games and, as a sub-category of this, experience gained through the playing of a specific FPS game which is then used in repeat playings of that game.

As regards socio-cultural experience of sound, there is a wide body of knowledge that humans acquire throughout their lives as to the mapping of parameters of sound to general or specific meanings.[3] As noted below, much of this is culturally-specific[4]

[2] A potential area for future research.

[3] It would be interesting to speculate as to how much of this acquisition is actually (or is developed from) an innate skill humans are born with as a result of evolution. An Ig Nobel prize winning essay suggests that the typical human reaction to the sound of fingernails scraping on a blackboard derives from a simian ancestor (Halpern, Blake, & Hillenbrand, 1986).

[4] Especially in the area of music.

and a good example is the semantics of alarms. The Western world is replete with sonic alarm signals ranging from clock alarms to telephone ringing to smoke and fire alarms to name but a few and their meanings are well enough known to the point where they have become conventions and dead metaphors. Such deliberately intrusive sounds, though, are unfamiliar to cultures lacking the requisite exposure.[5] This is not to say that these cultures are unfamiliar with the exclamatory function of sound; warning calls make use of this and many cultures will understand the alarm signals of other species. What it does mean, though, is that a person from one of these cultures is unlikely to be able to assess the specific meaning broadcast by the obtrusiveness of mobile telephone ring signals. It is a fair assumption to make that FPS games are designed and (designed to be) played within the one culture. This, if there is such a thing, is the First World culture which may be described, in this context, as that culture which has assimilated Western sonic conventions.[6] Of interest here, is that sonic conventions brought to the interpretation and use of FPS game acoustic ecologies are often conventions from another media.[7] This is particularly the case with cinema and, as suggested in chapter 3 and explained further in chapter 5, applies not only to game players but also to game sound designers.

It is often the case that dedicated FPS gamers, while they may have their favourites, play several such games.[8] Such games are designed within the First World culture utilizing its sonic conventions and the practised FPS player will notice a common sonic language. For example, the sound of footsteps indicates character movement, 'realism' FPS games will typically use the recorded sounds of real-world weapons and team communication speech samples are usually processed to sound like military radio messages.[9] This common FPS game sound language, the wide use of sonic convention, makes it easier for the seasoned FPS gamer to engage with the acoustic ecology.

Particularly at first exposure, the process of playing a game is also a process of learning the game and of training in the rules and conventions of the game not least of which is gaining an understanding of the sonic language of the game. Part of this involves deductions of sonic meaning from real-world experience (as in the case of the footsteps above) and thus makes use of socio-cultural experience. But part of it also involves the learning of meaning for previously unheard sounds and groups of sounds. This is particularly the case for FPS games, such as *Quake III Arena*, which are set in fantastical game worlds and so are more free to use fantastical sounds such as a variety of sounds indicating the spawning or use of powerups like quad damage or the teleporter. It is also the case, though, for the 'realism' FPS game such as *Urban Terror* which, although most of its audio samples are recording of real-world equivalences, does have sounds the meaning of which will be unknown to the *Urban Terror* novice. An example would be the musical sound indicating the taking of a flag in capture the flag mode. Learning the meaning of this sound leads to

---

5 A variety of peoples throughout Africa, for example.

6 This is an unsatisfactory term for a number of reasons and so is used here merely for convenience.

7 In fact, with *Urban Terror* (Silicon Ice, 2005), the weapon audio samples were borrowed from a Hollywood sound designer as noted in chapter 3.

8 In my case, my favourites are *Quake III Arena* (id Software, 1999) and *Urban Terror* but I do also play other FPS games such as *Return to Castle Wolfenstein* (Gray Matter Studios & id Software, 2001), *Battlefield 1942* (Digital Illusions, 2002) and *Doom 3* (id Software, 2004).

9 Bookended by static and with a highly compressed frequency range.

the acquistion of experience and means that, in later playings of the game, this experience can be brought to bear to interpret that sound with immediacy.[10] This learning process is an important instance of a relationship between player and sound and one that has implications for player immersion within and participation in the FPS game acoustic ecology as discussed in chapter 7.

As explained below, FPS game sounds may be listened to in a variety of listening modes, may offer multiple affordances and may have different functions. Experience as to the meaning of sound in any of the three fields described in this section has an effect on which listening mode is utilized, which affordance is given priority and which function is attended to first. I return to the effects of experience in subsequent chapters where necessary but the importance of it in the perception of and engagement with the FPS game acoustic ecology should always be borne in mind.

4.2 Modes of listening

Sound theorists, particularly those working in the areas of electro-acoustic composition and film, have suggested that there are several modes of listening which listeners may unconsciously or consciously use. In other words, when presented with a sound, the choice is made as to which parameters and properties of the sound to listen to and this affects the meaning that, in the first instance, is derived from that sound. It is a reasonable assumption to make that FPS players also make use of these modes of listening in order to make sense of the game world and, therefore, in order to participate in the game's acoustic ecology.[11] An FPS sound, therefore, may be approached through a mode of listening that prioritizes its locational properties (those sonic parameters that indicate the position of the sound in the game world relative to the player) and this information may then be used to navigate through a game level or to contextualize oneself in relation to game objects (including characters) which may or may not be displayed on the screen. Or it may be that the sound is assessed semantically and that, from the meaning thus derived, information about the game status, such as the relative scores of two teams, may be gleaned. In this section, I investigate existing modes of listening, introduce a new mode of listening as used in FPS games and explore the implications of the shifting of mode (whereby the player utilizes a different mode of listening after having gained experience in assessing the meaning of a particular sound).

In terms of how the audience listens to film sound, Chion (1994), developing the ideas of the composer Pierre Schaeffer, suggests that there are three modes of listening: *causal listening*, where the listener attempts to gather information about the sound source; *semantic listening*, where the listener utilizes a (semiotic) code to interpret (the meaning of) the sound, and *reduced listening*, where the listener perceives and appreciates the sound *sui generis* without reference to cause or meaning (pp.25—34).[12] Where the sound source is visible, causal listening is often backed up by visual confirmation when synchronising images on-screen to audio

---

[10] In some cases, as Schott and Kambouri (2003) have noted, this experience may place the player in the straitjacket of a player-developed game strategy compared to the more exploratory approach of the novice player.

[11] Future research may usefully and experimentally investigate the modes of listening that FPS players employ in order to provide an empirical basis for this assumption.

[12] These modes may be used simultaneously or the listener may switch focus between them.

originating from transducers.[13]

Game sound scholars have touched on these modes of listening by generally suggesting that the reduced mode of listening is probably little used when playing a game (Stockburger, 2003). From personal experience (and in the absence of any published research in the area), beyond first novelty it is unlikely that any FPS player consciously listens for aesthetic pleasure to the structure and grain of the sound (any who do, do not survive long in the game world). It would make sense to argue that causal listening, and particularly semantic listening, are of more importance to players of FPS games who depend on initially causally determinable audio cues to semantically distinguish threat from death-dealing opportunity, particularly when the source of the sound is not visible. As in this example here, it is likely that players utilize causal listening first (*what is causing that sound?*) followed by semantic listening (*now that I know what's causing it, what's the significance of that sound for me?*).

Friberg and Gärdenfors, describing the sound modeling of a game for visually-impaired players, make the statement (with no further elaboration) that mainstream games (that is, digital audiovisual games as opposed to non-digital or digital audio-only games), while making use of causal[14] and semantic listening modes, make little use of the reduced listening mode. The authors suggest that a greater use of this last mode would make the game experience more immersive and open-ended (Friberg & Gärdenfors, 2004). Their paper modifies Scott McCloud's (1993) semantic model for cartoons to a semantic model for sound (I presented their model as *Figure 2.2* in chapter 2):

While the model in *Figure 2.2* serves a purpose in highlighting some aspects of sound potentially useful for further inquiry, there are several problems with it. There seems to be a clear-cut distinction between speech sounds and 'authentic sounds' with the former only being listened to semantically. This distinction (with the addition of being buffered by 'cartoonised sounds') is puzzling since both are presumably indexical in nature as opposed to the iconic 'cartoonised sounds'. Furthermore, the authors state that "[s]emantic listening is used when understanding auditory codes such as speech or Morse code" (Friberg & Gärdenfors, 2004) and, in common with Chion's application of the mode (p.28), implicitly exclude any form of non-linguistic sound from having semantic properties. According to these authors, for any sound that is not derived from language codes, semantic listening is a futile exercise since such a sound is devoid of meaning. As this chapter makes clear, I disagree with this; an example of my stance on this matter would be the semantics of alarms where, for instance, in a given culture a fire alarm has a very specific denotation and potentially a variety of similar connotations resulting from experience, familiarity and, to some extent, the physical structure of the sound.[15] Additionally, the model gives the impression that causal listening has little if any utility when listening to speech in a game. Where speech does occur in FPS games, it seems logical that the cause of speech is as important as the meaning of that speech. As a simple example, imagine the different outcomes of obeying a request to *meet me in the pump room!*

---

[13] Note that I very deliberately state that, in effect, it is the ear that guides the eye and this is a theme I return to throughout this thesis.

[14] Which the authors unfortunately refer to as the *casual* listening mode.

[15] The mere fact that it is possible to use semiotic terminology to describe non-linguistic sounds points to the possibility that such sounds may be attended to with the semantic listening mode.

where the origin (cause) of the speech is not known to be friend or foe.

Finally, depending on player experience and familiarity with any game, for any particular sound the mode of listening is likely to change. Although some sounds are dynamic in nature and can appear at different times and different locations within the game play (for example other characters' sounds in a multiplayer game), others tend to be static and may, for example, be associated with a particular location. The player knows that at a particular location in the game, a certain sound always becomes audible (Stockburger's 'zone sound') and can be followed to its source. At first playing of the game, this sound may be puzzling. However, because the sound follows normal acoustic behaviour (the inverse of the inverse square law so to speak) in becoming louder the closer to the source the listener is, the player can follow the sound to its source where, for example, it may be discovered that the noise comes from machinery in a pump room.

It may very well be that the player has utilized causal listening to determine that the sound is mechanical in nature (cyclical, metallic properties for example) and it may well be that the player extends this causal listening further to rationalize that perhaps that mechanical sound is issuing from behind that door labelled 'Pump Room' (perhaps inside which there is an ammunition store or the aforementioned friend or foe). However, this listening and reasoning is only used until the player has become familiarized with the sound and its cause (a particular artefact in the game). Thereafter, the player has the possibility of listening to and utilizing that sound for locational purposes — an audio beacon. This mode of listening is neither reduced nor semantic nor causal listening but something I shall term *navigational* listening.[16]

Sound theorists from the electro-acoustic composition and cinematic traditions have no use for the concept of using a sound for orientation in a 3-dimensional environment. Listening to recorded music or watching a film is an activity lacking in much physical, haptic interactivity;[17] in neither case is it possible for the listener to move within the musical or cinematic environment (that is, the filmic environment displayed on the screen) following a sound to its apparent source. This is only possible in real or virtual 3-dimensional environments, such as the acoustic ecology of the FPS game, and thus the requirement for the fourth mode of listening. In this case, the first listenings to the sound have been used to construct a representation or mental map of the game level (which may be a set of ordered images or a set of route instructions) within the player's mind through which the audio has helped in navigating.

Chion suggests that "[w]hen we listen acousmatically to recorded sounds it takes repeated hearings of a single sound to allow us gradually to stop attending to its cause and to more accurately perceive its inherent traits" (p.32). In other words, our listening shifts from causal listening to reduced listening. For FPS audio beacons of the type discussed above, the player's listening shifts from causal listening to navigational listening and navigation of the game level is of great importance in FPS

---

16 The term *navigational* listening is preferred to *mapped* listening because, although referring to memory and stored cues, navigation suggests action and takes place in the present being a real-time activity. Furthermore, this listening process and navigation is what leads to the construction of accurate and navigable mental maps rather than maps, which like those of some of the more home-bound but imaginative ancient geographers, cartographers and compilers of mediæval Baedekers, owe more to supposition and fancy.

17 Certainly, once the music or film has started, its continued playing requires no further physical input from the listener or spectator.

games. These issues will be discussed further in chapter 6 which deals with sound and spaces in the FPS game acoustic ecology.

As suggested by the game sound theorists discussed above, the FPS player makes use of a variety of modes of listening and I propose that the prioritization of these modes depends upon context (as that context relates to image, event and FPS game spaces, I discuss this in subsequent chapters), experience and prior knowledge of the game or genre. Reduced listening is little used beyond initial exposure to the game's sounds. Other modes (causal and semantic) are of more use in contextualizing the player within the FPS game acoustic ecology. Navigational listening is is used by players to orientate themselves within the game world whereby some sounds may be used as audio beacons. The identification of such a listening mode is an important distinguishing feature of the FPS game sonic environment when compared to the environment of electro-acoustic music and the sound of cinema in which theoretical fields the initial three modes of listening were first identified and subsequently theorized. The significance of causal listening will be explored further in chapter 5 and navigational listening in chapter 6.

4.3 Affordance

The FPS game acoustic ecology offers the player the opportunity to immerse himself in the ecology and to participate in it. Furthermore, each sound event or sound object provides opportunity for a range of responses comprising interpretations that are usually followed by actions. Such opportunities are termed affordances in the literature. Any sound object may have multiple affordances the prioritization of which is often subject to prior experience, socio-cultural conditioning and convention and, importantly for the FPS game, the context in which the sound is heard. A study of the range of affordances offered by the FPS game acoustic ecology is, therefore, a prerequisite for the study of player immersion and the player experience that is part of chapter 7.

Gibson's definition of affordance has been summarized by Zhang (n.d) as:

> Affordances provided by the environment are what it offers, what it provides, what it furnishes, and what it invites [...] Affordances are holistic. What we perceive when we look at objects are their affordances, not their dimensions and properties [...] They can only be measured in ecology, but not in physics.

Originally introduced as an objective term referring to all action possibilities on the part of the object and within the capability of the subject, it was later refined into a subjective phenomenon by including the consequences of knowledge, belief systems and socio-cultural experience on the perception of the affordance (2004-2006). To Gibson (1966), a weapon, for example, has a set of objective and invariant affordances to a human being. However, the modern definition of the term affordance accounts for what is likely to happen when the subject (with his own knowledge and experience) perceives an object with consequent affordances.

The later subjective definition accounts for the possibility that some cultures, societies or persons may be unfamiliar with that particular weapon and will therefore not perceive the same affordances, using it, perhaps, to dig in the ground with or to hang a hat on rather than fighting enemies with it. However, it also accounts for the possibility in FPS games that a repeat player, by now having had some experience of the sounds heard and having gained knowledge as to their significance, may be more attentive to certain sounds while sublimating the affordances offered by others as having lesser importance to the gameplay (perhaps in the case of environment sounds). Affordances, therefore, may mean different things to different people and, furthermore, for the one person, are dynamic; increasing knowledge of or experience in the use of an implement is likely to change the hierarchy of affordances offered by that implement such that the user's ultimate prioritized affordance is likely to closely, if not exactly, match the intended and designed-in affordance of that implement.

Affordances are prevalent in the FPS game (the whole acoustic ecology may be viewed as one which offers the affordance of immersion within and participation in the ecology and the entire game itself may be analyzed as a set of affordances). Gibson suggests that when we look at an object we see not its physical dimensions and other properties but its affordances (this concept is similar to the ways in which we perceive sounds as described by Gaver below). These affordances may be defined by the object's physical properties — a mortar and pestle set offers the affordance of crushing herbs and spices, of eating from the mortar or of using the pestle as a blunt weapon because it has a bowl, is hard and non-porous and has a pounding implement — but what we prioritize, assuming we have the requisite prior cultural experience and knowledge or, because we are able to provide the correct relationship between the implement and the physical, culinary context, is the first affordance. Similarly, affordances in FPS games are defined by the game engine; this is never seen or sensed in any way during the playing of the game, only the affordances it offers are perceived. An FPS game visual affordance may manifest itself in the shape of a gun, a stairway, a river, a motorboat, aeroplane or any of the other objects and constructions a player may interact with.

Likewise, sound has its affordances too because sound has value and meaning that can be perceived. As with physical or in-game visual objects' affordances, it can be said that when a sound is heard, what is perceived is not the physical properties of sound (frequency and amplitude for example) but its affordances, its use value to the perceiver, what is offered and what is furnished. So, in an FPS game, footsteps, shooting, alarms and audio beacons all offer affordances to the perceptive player and these affordances are of the subjective variety as they potentially have different affordances to different players (footsteps will have varying affordances for players on different teams, for example, and prior experiences and different socio-cultural backgrounds play their part too).

FPS game affordances are more constricted and limited than affordances in reality. Thus, in reality, a gun may have any number of affordances; the likely affordances being shooting, threatening, bludgeoning with the stock and the least likely affordances (when cognitive dissonance occurs) being anything the user can think of from using the gun as a walking stick, a prop for a hat to using the barrel as a flower holder.[18] In an FPS game though, the uses to which a gun may be put are severely restricted by the game engine and therefore the only affordance a gun is likely to

[18] The cognitive dissonance that occurs in the latter case is a probable reason why this Vietnam War-era image is such a powerful icon.

offer in the FPS game is the affordance of shooting.[19]

This last is an interesting situation and is a point of difference between real-world affordances and virtual world affordances. Virtual analogues of real-world objects tend to offer fewer affordances (although that restricted set is usually drawn from the larger set offered by the latter object). Such affordance restrictions do exist in reality — the aforementioned mortar and pestle may not be used for travelling purposes[20] — but tend to have a larger set of possibilities than that offered by the FPS game engine. The process of learning an FPS game (or the operation of anything for that matter) may be described as a process of learning what is afforded by the game objects (including sound), of prioritizing affordances and, therefore, of learning to make efficient use of such affordances.

The initial use made of affordances offered may be summed up as *awareness* (and this itself may be termed an affordance) and there are two types of awareness in the FPS game which are afforded by sound. Firstly, there is the awareness that one is within an acoustic ecology and sounds offering this type of affordance are those that may be described as ambient sounds, indicating resonating space (that is, that the game world has (virtual) volume) and paraspaces such as location and temporal period, and other sounds that, in their processing, offer the same spatial affordances. As an example, a gunshot sound may have been processed to include reverberation[21] and may have directional and depth qualities, all helping to make the player aware of operating within a resonating space. But a particular gunshot sound also provides an awareness of paraspaces, such as historical period due to the material qualities of sound (the game is not set in the stone age or, because it is machine-gun fire, the action is set during or post-Boer War). Secondly, many FPS sounds provide an awareness that something is going on, that there is action occurring within the game whether that action is a result of the player's own input or results from other characters or other gameplay events. In the latter two cases particularly, such an awareness invites a response (which, typically in the FPS game, triggers a further sound itself ripe with affordances) and it is this ability to be responsively aware of the affordances offered by the game's sounds that is one of the keys to immersion and, particularly, participation in the game's acoustic ecology.

Affordance theory, then, offers one way to analyze the range of possibilities offered to players by sounds when developing the concept of the FPS game acoustic ecology. The sounds comprise a set of affordances and so the acoustic ecology would be an ecology of opportunity. Because the acoustic ecology requires the participation of players in order to be experienced as such (see chapters 7 and 9), these affordances are fundamental to the creation and maintenance of the acoustic ecology. The sonically aware player is able to contextualize himself within and participate in the game world through the affordances offered by the game's acoustic ecology. This contextualization is a process of immersion in and response to both the game's environment and the game's action and is dependent upon prior experience and socio-cultural background in its assessment of affordances. The

[19] Inventive players often discover other affordances which do not necessarily contravene the rules of the FPS game. For example, the player who artistically or literately tags a wall in the game with a spray of machine gun fire.

[20] Unless the mortar, propelled by the pestle, is capable of flight and one happens to be a Russian witch named Baba Yaga.

[21] Depending on the abilities of the game audio engine, this processing may be in real-time or it may have been applied prior to the game's packaging.

assessment of sonic affordances on the basis of experience leads to a prioritization of affordances and the affordance at the top of the hierarchy is likely to be the one that enables a more immediate engagement with the game world and the gameplay.

4.4 Auditory icon design

Having described the listener's primary engagement with sound through modes of listening and having suggested that any one sound has a variety of affordances, the prioritization of which is dependent upon context and experience, I now turn my attention to the derivation of specific meaning from sound. Much of this has been theorized in literature which deals with the semiotics of sound. In particular, the theory of auditory icon design proves beneficial to an understanding of this and so is discussed in this section. I suggest that sounds in the FPS game may be described as, and classified as, a variety of auditory icon types. This method proves to be a useful preliminary exercise prior to subsequent chapters where the meaning of sound is investigated in the context of FPS game elements such as images, actions and spaces. In chapter 9, I use the FPS game *Urban Terror* to illustrate the hypothesis of the FPS game acoustic ecology. Here, I use *Quake III Arena* in order to widen the possible applications of the ideas expressed here to an FPS game which demonstrates less realism than the former.

The design of auditory icons is concerned with encoding meaning in sound for use in audio or audiovisual displays. Such an icon consists of a sound object (which is usually a pre-recorded audio sample but may also be synthesized in real-time) which is sounded whenever a particular event occurs in the digital system of which it forms part. The intent is that it indicates that the event has taken place and common examples include the Microsoft Windows audio sample that is played whenever a file or directory is dumped in the recycle bin or the ringing of a telephone. Auditory icons, therefore, can range from sounds that are similar to real-world analogues of the digital action to more abstract icons. The former often require socio-cultural knowledge or experience of the real-world equivalent while the latter require prior experience of the auditory icon in context. Acoustic ecologies (which I suggest may be modeled technically as sonification systems— see section 4.5 and chapter 9) are comprised of sounds that provide meaning to the listener; in the FPS game's acoustic ecology, the audio samples have been recorded and designed to impart specific meanings in similar manner to the design of audio icons. Furthermore, like audio icons in an audiovisual system, these audio samples often work together with image to amplify any such meaning. A study of auditory icon design and the methods for the encoding of meaning within them, therefore, is likely to provide useful insights into the similar use of audio samples within the FPS game's acoustic ecology.

For Blattner, Sumikawa and Greenberg (1989), earcons[22] are "nonverbal audio messages used in the user-computer interface to provide information to the user about some computer object, operation, or interaction" (p.13) and they define representational earcons (sound that is a natural representation of the computer object or operation being described) and abstract earcons (having only a symbolic relationship to the real-world object or operation being depicted in the user-computer interface). All such abstract earcons, according to their system, share the property

22 The term they use for auditory icon, a pun on 'icon'.

that their meanings must be learned not only as separate sign elements but as sign elements that can be combined, either in parallel or in series, to produce new complex earcons with accordingly new and complex interpretations. Pitched sounds and unpitched sounds, a variety of timbres, shades of soft and loud, diminuendi and crescendi, rhythm, tempo, melody and harmony are all used within a system that the authors note should be ideally designed by a professional composer (Blattner *et al.*, 1989, p.23). I am unaware of any human-computer interface that uses this abstract and *musical* earcon system to any great degree beyond the use of simple, uncombined earcons and, given the complexity of such a system, not only to design but also to learn and interpret, the reader will not be surprised to learn that I dismiss this stunningly Laputian endeavour as having little or no relevance to FPS game sound.

Of more pertinance to this work is Gaver's (1986) approach. In particular, his description of *nomic* auditory icons proves to be of use in the classification and analysis of FPS game sounds which are recordings of real-world objects and actions equivalent to those shown in-game. Writing earlier than Blattner, Sumikawa and Greenberg, he describes the design and implementation of auditory icons that bear similarities to the latter authors' representational earcons (although he himself does not use the term earcon). For Gaver, "an auditory icon is a sound that provides information about an event that represents desired data" (p.168). His approach is strictly representational in that there is a direct correspondence between the sound heard and the object or operation it represents. As Gaver explains, "[i]nstead of using dimensions of sound to stand for dimensions of the data [as in the abstract earcon], dimensions of the sound's source are used" (p.168). In other words, Gaver's technique is one which uses parameters such as the material, size and density of the object or the qualities of the action to transfer meaning rather than parameters of sound such as frequency and amplitude. This, as explained in section 4.5, is similar to the use of sonification techniques for the apprehension of non-audio data. The importance here is that sounds in the FPS acoustic ecology directly relate to the qualities of actions or parameters of objects in the game.

According to Gaver, symbolic auditory icons (and it is not clear whether he in fact does regard these as true auditory icons) have an arbitrary mapping between the data and its representation (p.170). These icons make use of parameters of sound such as pitch and amplitude and are therefore analogous to the abstract earcons of Blattner, Sumikawa and Greenberg. Because this mapping is arbitrary, such icons must be learned and therefore are not as direct and immediate in the transfer of information to the user as other forms of auditory icon. The exceptions being the case where the sound of the icon is culturally ingrained such as the ringing of a telephone bell or the wailing of an ambulance siren.

At the other extreme are what Gaver terms *nomic* mappings in which there is a (partly) causal relationship between the sound of the icon and the object or operation it represents (p.170). These are typically recordings of the real-world objects or operations but may also be caricatures in which salient sound features are emphasized at the expense of others. A one-to-one nomic mapping for sound is a recording of an object or operation with minimal audio processing and is analogous to an untouched photograph; a caricature nomic auditory icon representing a door shutting may emphasize the slam of the door on the door frame while ignoring or attenuating the squeeking of hinges or click of the lock. In a later article, Gaver

(1993) demonstrates the use of synthesis rather than recording to create nomic mappings in which the synthesis is modeled not by sound dimensions such as frequency and amplitude but by salient sound characteristics such as a scraping sound, a bouncing sound or a breaking sound for example; such characteristics have been identified and isolated by the work of Warren and Verbrugge (1984) and form the basis of Gaver's demonstration.[23]

Where nomic mappings are not possible, Gaver suggests the use of metaphorical auditory icons in which the mapping is of two types (p.170). Structural mappings are based on similarities of structure; Gaver (1986) uses the example of descending pitch to represent a falling action (see Curtiss' isomorphic sounds discussed in chapter 5). Metonymic auditory icons are described by Gaver as those "in which a feature is used to represent the whole" (p.170); the example given is a hiss representing a snake (a recording of a snake hiss would be nomic so here the hiss could be the sound of steam escaping under pressure from a valve).[24] Metaphorical mappings at the extremes tend towards symbolic for weak metaphors and nomic for strong metaphors.

The advantage in using nomic auditory icons (or strong metaphorical auditory icons) lies in their greater articulatory directness compared to weak metaphorical or symbolic auditory icons (with the aforementioned exceptions of culturally ingrained auditory icons such as telephone bells and sirens). The greater the articulatory directness of an auditory icon, the less effort or training is required in learning the meaning of it. Symbolic relationships must be learned and often must be learned specifically for the user-computer interface at hand. For Gaver, a nomic auditory icon is "a natural and intuitive way to represent dimensional data and to represent conceptual objects in a computer system" because humans are accustomed from early childhood to making causal connections between a sound and a source based on the data that the sound carries with it (p.176).[25]

There are few symbolic auditory icons in *Quake III Arena* and most such sounds are to do with the menu interface, that is, they are nondiegetic. However, those that are diegetic symbolic auditory icons include a variety of sounds to indicate the status of the two flags in capture the flag configuration. Thus, when a flag has been taken, is dropped or has been returned to its base, these actions are accompanied by the sound of a power chord on an electric guitar or by a similar synthesized sound.[26] There are likewise some metaphorical auditory icons in *Quake III Arena* in which the sound has some structural similarities to the object or event depicted — the sound of the personal teleporter in action is an example.

The majority of sounds in *Quake III Arena* are nomic or representational having either a one-to-one mapping (analogous to a representational photograph) or displaying

---

[23] Coward and Stevens (2004) have provided experimental data using Gaver's synthesis models in which listeners were able to predict the source object's properties with an acceptable degree of accuracy.

[24] In the sense that Gaver uses the term, a *metonymic* auditory icon would be better termed a *synecdochal* auditory icon.

[25] In an argument against the assumed universality of icons (both visual and auditory), Gaver's and Blattner, Sumikawa and Greenberg's systems have been criticized by Familant and Detweiler (1993). As suggested in section 4.1, player experience and prior knowledge play a part in assessing the meaning of sound so the latter authors' assessment is likely to be the correct one.

[26] The synthesized sound may in fact simply be an electric guitar chord that has been so heavily processed as to be unrecognizable as an electric guitar. While not therefore synthetic, it is processed to sound like a synthetic sound.

some aspect of caricature in which salient features of the object or action being represented are emphasized at the expense of others. Because every single recording in *Quake III Arena* has been processed in some manner, a strict reading would state that there are no one-to-one nomic auditory icons in *Quake III Arena.* Instead, it is more accurate to state that all nomic auditory icons are of the caricature variety with the degree of caricature dependent upon the degree of processing applied. For some of the more human characters, the footstep audio samples are close to a one-to-one nomic representation. Other sounds, such as some of the speech announcements (bass frequencies enhanced and some distortion applied) or the shotgun sound (bass frequencies enhanced) are more like caricature representations in which an audio recording has been heavily processed to emphasize some important dimension of the source object or action, or in which a sound has been created specifically for the purposes of caricature.

Not only are most sounds in *Quake III Arena* nomic, but it is the case that any symbolic and metaphorical auditory icons are used for actions and objects within the game for which there exist no real-world equivalents. Thus the spawning and use of a variety of power-ups (quad damage, mega health or regeneration for example) or the use of fantastical objects (such as the teleporter and jump pads) are accompanied by either symbolic or metaphorical auditory icons. Conversely, in-game objects and actions for which there are real-world equivalents are represented by nomic icons existing on a scale between caricature mapping and somewhat less than one-to-one mapping. In FPS games that attempt a more all-encompassing illusion of reality than *Quake III Arena* (such as *Battlefield 1942* or *Urban Terror*) there are very few, if any, in-game, diegetic symbolic or metaphorical auditory icons. Even the auditory representation of a sinking ship in *Battlefield 1942*, presumably a sound difficult if not impractical to record, is convincingly represented by a caricature nomic auditory icon consisting of the sounds of a large metal object under stress.[27] Therefore, the degree of the simulation of reality in an FPS game can, to some extent, be assessed by the proportion of nomic sounds to non-nomic sounds.

Because an FPS game comprises a visual and auditory interface, the sounds in an FPS game can be considered as auditory icons. The purpose of the FPS auditory interface is to provide feedback about the course of the game to the player (except in the case of add-ons such as voice-over-internet which enables team members to communicate bilaterally by voice). It is the means by which the acoustic ecology is engaged with and participated in (which, with other factors, leads to participation in the gameplay). The engagement and participation is enabled by ascribing meaning to the sounds heard; this is a process that originates at the sound design stage but, in order for the intended meaning to be transmitted, the meaning of FPS game sounds must either be learnt in context (or in a similar situation) or both sound designer and player must have similar socio-cultural backgrounds.[28] The analytical techniques of auditory icon design are, therefore, likely to prove valuable when assessing the derivation of meaning within the FPS game's acoustic ecology which itself may be technically modeled as a sonification system; the subject of the next section.

---

[27] It might be argued that this is a synecdochal metaphorical auditory icon but I suggest that it displays enough of the salient features of the represented object and action to be classed as a caricature.

[28] A spurious argument for coca-colonization.

4.5 Sonification

Sonification theories provide techniques for assessing how non-audio data, such as gameplay events and player input, may be expressed in sound and therefore provide a means to map the non-audio event or player input to the meaning that players perceive in sound. In other words, as a theory, sonification provides a bridge between action and perception and sound becomes a medium or language through which the game engine communicates to the players and through which players communicate with each other — relationships that are potentially at the core of the acoustic ecology. But sonification also provides a possible model, in part at least, for the practical modeling of the FPS game's acoustic ecology (see chapter 9). That is, it helps to describe how that acoustic ecology is technically constructed prior to the process of player immersion and participation within it. The FPS game acoustic ecology is a type of space that is at one and the same time a resonating space and a set of paraspaces (chapter 6). In order for the player to become an integral component of the resonating space, that space must first be conjured into existence and it is here, in understanding the process by which this takes place, that sonification concepts prove most useful.

Kramer *et al.* (n.d.) provide a definition of sonification as the use of a sound generator to transform non-audio data into sound in order to facilitate, or perhaps provide new, understanding of that data (p.3). Sonification techniques, therefore, are intimately concerned with expressing meaning in sound. The FPS game engine may be viewed as a sonification system in that it translates non-audio data (the game status or player actions, for example) into sound through the agency of the computer's or console's audio hardware thereby providing sonically interpretable data to the player. This section, then, discusses the relevance of sonification concepts to the hypothesis of the FPS game acoustic ecology and provides a method for modeling the game engine as a sonification system.

The purpose of sonification is to monitor and comprehend data which might otherwise, and in another form, be difficult to monitor and comprehend. The transferral of data to the audio domain allows the specialized skills and abilities of the auditory system to be brought to bear on that data in an attempt to discern meaning from it. As an example, the ears are very finely attuned to minute temporal and frequency changes, more so than the eyes. Furthermore, and this is important for monitoring purposes, the ears, unlike the eyes, cannot be shut down (there are no earlids) and need not be oriented in any particular direction to sense sound.

Kramer further defines two other terms related to sonification. Firstly, audification (sometimes referred to as 0th order sonification) is "[a] direct translation of a data waveform to the audible domain for purposes of monitoring and comprehension" (Kramer, 1994, p.186). As it is a direct translation, the data must exist in the form of an analogue or digital waveform simply requiring amplification, transduction and little other processing in order to be heard. Examples of this include wavetables stored in synthesizers or the grooves in vinyl records. Secondly, audiation (sometimes called auralization or realistic sonic spatialization) is the formation of "an imagined auditory image" and requires cognitive functions such as memory and recall (Kramer, 1994, p.188).

Audiation is classed with sonification by analogy. Whereas sonification (and this includes audification) technologically transforms physical data into sound sensation, audiation occurs solely within the mind as an expression of the inner ear of the mind. It may occur when the visual system is confronted with an image of an animal, a dog for example, in which case the inner ear provides the appropriate bark based on prior experience or expectation (the viewer may not have seen or heard that particular type of dog before but recognizes that it is a picture of a lap dog and therefore provides the appropriate audiated yap). Or the audiation process may involve the formation of more imaginative sonic mental imagery if confronted by the visual image of something completely unrecognized or perhaps even non-existent.[29]

Audiation has a role in the perception of presence or immersion within virtual environments such as that found in FPS games. As Fencott (1999) suggests, the stimuli or, in this case, the accurate representation of real-world sounds within the game, are of less importance than the audiation and visualization processes players undergo because the sense of presence or immersion in the game is a mental construct resulting from perception rather than sensation. Audiation may also be viewed as a motivating component of Ermi and Mäyrä's imaginative immersion criterion (as discussed in chapter 7) and I would suggest if a player, new to the FPS game in question, audiates a sound for a particular object and this audiated sound is then closely matched when the game object actually sounds, that this process of expectation and confirmation contributes greatly to the perception of immersion in the FPS game by granting the player a sense of partaking in or of ownership of the imaginative construction of the game world. This is similar in many respects to the comments made earlier in this chapter about the use of perceived affordances and the causality of sound in constructing the perception of a believable, and therefore usable, game world.

The system of mapping of non-audio parameters to audio parameters, while it may be entirely arbitrary, is, if it is to be used by as wide a circle of people as possible, typically based on metaphor and affect; ideally, commonly understood metaphor and commonly experienced affect. Kramer outlines some mappings based on metaphor in which the non-audio parameter of *more* (of anything — population density, incidence of tooth decay for instance (may be mapped to audio parameters such as louder (larger sound objects, more sound intensity), brighter (more partials, more high-frequency energy), faster (more sound in the same time frame) or higher pitch (greater frequency, more cycles per second) for example. Affective asssociations relate to the feelings aroused by sound. For example, desirable and undesirable qualities (perhaps lesser and greater incidence of tooth decay) in the non-audio data set are mapped to pleasant or ugly qualities of (in this case musical) sound such as harmony and disharmony or in tune and out of tune (Kramer, 1994, pp.212—217).

Both metaphorical and affective associations such as these are subjective and depend for their recognition (and consequent reverse mapping back to the non-audio parameter) upon a number of factors not least of which are training and cultural or societal experience. Although metaphorical association between *more* of a property and increasing sound intensity may be viewed as a global metaphor (it is highly likely that whatever the culture, the connection between more sounding objects and

---

29 I imagine science fiction and horror film sound designers must gain enormous pleasure from giving voice to the film's creatures deriving it from some audiation that is based upon expectation (a big creature requires a big sound), role and the character of the creature within the film and upon prior film practice. Sonification literature has no term for or description of this auditory reification but it is a form of sonification.

greater overall loudness of sound will be recognized), the association between *more* and higher pitch may require some prior training. This may be part of the general training or schooling of a culture (relating increases in pitch to more cycles per second or rising vertical displacement on a musical stave) or may be training specifically for the sonification system in question.[30] Affective associations run the risk of being even more culturally specific or dependent upon training, especially when the non-audio parameters are mapped to musical parameters. Different societies have different musical aesthetics and consequently varying notions as to what is a consonance and what is a dissonance, what is in tune and what is out of tune. However, affective associations do have benefit in creating atmospheres in the game.[31] Mysterious, acousmatic sounds prove to be particularly effective in creating an atmosphere of dread especially if the player visualizes a horrible monster lurking in the shadows (see chapter 5 for a discussion of such sounds and the term 'acousmatic').[32]

A simple alarm system, for example a fire alarm in a building, uses sonification by mapping (depending on the system) the density of particulates in the air or the ambient temperature to a complete audio event (the alarm sounding). This mapping from a non-audio domain to the audio domain is crude but effective and, in the Western world at least, there is a wide knowledge of what constitutes an audio alarm despite the myriad of audio alarm forms. Thus, the smoke alarm in this case, is likely to lead to (or at least is designed to lead to) one immediate result which is the evacuation of the building. To semioticize the sonification at a first simple level, the audio alarm is the signifier and what is signified is 'danger'. At a second level, 'danger' may be the denotation and 'evacuate' would be one of the desired connotations. *Battlefield 1942* is one of many FPS games to use alarms (to signal to a team, for example, that a bombed factory needs urgent repair) and game status sounds indicating the taking of a flag, in those FPS games such as *Quake III Arena* which can be played in capture the flag mode, function as team alarms.

A more complex sonification system might be what is termed a 'network auralizer' — a sonification system that is used to monitor the status of a complex computer network.[33] In such a system, a pleasant and soothing continuous loop of sound might be an indication that the network is functioning and functioning, further, within expected norms. When a user logs on or off, non-threatening tones rise above the ambient sound (perhaps the physical location in the network of that activated host computer might be indicated by positioning the sound within the loudspeaker system or by a more complex class of tones). Events of importance, such as server crashes or network intrusions would be indicated by more strident tones (alarms in fact).

It may be possible to analyze components of the FPS game acoustic ecology[34] as forming a sonification system, similar to the one described above, which makes use of audification, sonification and audiation. By this analogy, audification is represented by the playback of audio samples. FPS run and gun games typically

---

30 There are some analogies here to the training required to interprete visual (re-)imaging of objects in which unseen parameters are mapped to visual parameters — the mapping of temperature or the Earth's ozone layer thickness to colour as found in systems such as NASA's Earth Observing System for example.

31 Atmospheres in the FPS game acoustic ecology are discussed further in chapter 8.

32 See Halpern *et al.*, (1986).

33 See, for example, Peep (n.d.).

34 Or, for that matter, any acoustic ecology including natural ecologies.

use hundreds of such samples which may be pre-synthesized or may be recordings of real-world events such as samples of various weapon shots or reloading sounds as used in the game. Because these audio samples are not sounds but are representations of the original sound stored as bits on the game medium (either removable media or computer hard drives), in order to be heard they must be passed through the sound generators that are part of the game system hardware; this is the process of audification or 0th order sonification.

Sonification, other than 0th order, is represented by the mapping of game events to audio samples which are then translated by the game system sound generators into the sonic domain in order to be heard. These game events may be initiated by players or bots as they partake in the gameplay and, in the case of players, interact with the game world. Examples would be the firing of weapons, movement, the collection of various game items or more significant and global events such as the capture of a flag. They may also be initiated by the game engine; game status signals usually decided before gameplay starts (such as the length of the game which, in some FPS games, is illustrated by sounds indicating how much longer remains to play). Environment, background sounds that are heard globally and that cycle throughout gameplay are strictly classed as audificated sounds, because they are not sounded by game events, but other environment sounds may be classed as sonificated sounds if, for example, they only sound when a player moves into a particular location in the level.

Audiation is represented by the player's expectations of what an object in the game should sound like (from previous game experience or cultural knowledge). In this model of FPS sonification, audiation may occur when a player sees a game object that is (currently) silent or when a player is listening out for a particular game event that is signalled by a sound. This might be the use of a power-up as in *Quake III Arena* or it might be the approach of an enemy into the team's stronghold. In both of these instances, the player is able to audiate the sound they are listening for based on prior experience of the hearing of that sound and the matching of it to a particular in-game event;[35] a process of categorization and memory involving perception and cognition. Playing the game again, the player expects to hear these sounds once more and, more importantly, will therefore be especially attentive to their occurrence; Bruner (1957) suggests that the expectancy of encountering events or objects in any context "preactivates a related array of categories" leading to a heightened state of perceptual readiness (p.137). Knowledge of and experience in the game, therefore, will have an effect on audiation possibilities as the player categorizes and prioritizes the various affordances offered by the game.

In this section, I have suggested that, through a process of active sonification of the game's components, by player input and by the process of player audiation, the technical processes of the game's acoustic ecology may be explained. That is, the sounding of audio samples and the potential for deriving meaning from those audio samples is a joint venture on the part of game engine and player. Additionally, sonification may also provide the interactive framework that is required to channel relationships between players and between players and the game engine. This notion is returned to in chapter 9.

---

[35] Or, possibly, when confronted with a particular visual object in the game, deducing its likely sound from real-world experience.

4.6 Conclusion

Sound is not only an experience of sensation, whereby the physiology of the human auditory system responds to auditory stimuli (or perturbations) at a first level, but it is also a process of perception and cognition where the hearer's prior experience and knowledge is brought into play in assessing the qualities and importance or meaning of the sound. Additionally, at first exposure to the FPS game in question (or to any new acoustic ecology), it is a process of training. The sound heard in the FPS game is therefore an individual experience and, while this sound may be analyzed on the basis of sensory data (such as frequency, amplitude and time), it is more interesting and valuable to analyze it on the basis of the cognition and, particularly, perception of sound. This is because the method by which meaning in sound is derived and the consequent meaning that is given to sound are key to the use of sound in understanding the FPS game world. This understanding, I suggest, proceeds from the immersion of the player within the game's acoustic ecology (see chapter 7).

Players utilize various modes of listening to aid in engagement with the FPS game acoustic ecology and, given the 'hunter and the hunted' premise of the FPS game, it is likely that the causal listening mode is one with a high priority while the reduced listening mode is little used. The semantic listening mode is also used in some cases (particularly in the case of speech) and I have identified a fourth listening mode which I term navigational listening and that aids in the orientation of the player within the illusory 3-dimensionality of the game world depicted on the screen. Sounds in the FPS game acoustic ecology may also be assessed in terms of their affordances and, indeed, the entire acoustic ecology may be viewed as a set of affordances or as having one main affordance which serves to draw the player into the FPS game world.

Theories behind the design of auditory icons and on the functions of sound provide useful ideas as to how to design meaning in sound and how listeners perceive meaning in sound. Sounds in an acoustic ecology either have meaning or they do not. If there are more of the former, the engagement of the listener with that ecology is all the quicker. The meaning of sounds can be learned and so the process of the player's engagement with an FPS game acoustic ecology is also a process of ongoing learning. This is also likely to be the case with an FPS game acoustic ecology and the more easily and quickly understood sounds the game designers include, coupled with a consistency in the use of those sounds, the greater the ease with which a player may be immersed within and participate in the acoustic ecology and therefore the game world. Conversely, challenge-based immersion (see chapter 7) requires that the meaning and potential uses of some sounds should take longer to learn.

Sonification has been discussed here because the techniques of sonification assume that it is possible to translate non-audio data into audio data the sounding of which either facilitates the understanding of such non-audio data or enables the apprehension of obscured or hidden meaning in that data. As such, sonification has the potential to serve as a technical model explaining the translation of non-audio data in the FPS game (such as game play events and the representation of spaces in the game world) to sound from which the player may derive meaning. If this is a valid model, it also helps to explain some of the relationships between players and the game engine as mediated through sound. I return to and expand upon this potential use of sonification concepts in chapter 9.

The preliminary conceptual framework was initiated in chapters 2 and 3 through a literature survey and through an investigation of the design and production of FPS game sound and the organization of audio samples in the game code. It has now been enlarged both in its conceptual language and in its taxonomy. The original three modes of listening have been supplemented by a fourth mode (navigational listening) to account for the 3-dimensionality of the FPS game world and the opportunities for movement within that environment. Furthermore, sounds in the FPS game furnish the player with affordances and the hierarchical prioritization of such sonic affordances represents a fundamental level of engagement with FPS game sound through which the player may derive basic meaning. Additionally, sonification may serve as a paradigm with which to model both the technical creation of the FPS game acoustic ecology and to model the sonic relational framework between players and between players and the FPS game engine. Expanding the taxonomy, FPS game audio samples may be classified as symbolic or nomic auditory icons and may be assigned a range of functions (deictic, exclamatory, simile, metaphoric and onomatopoeic). Accordingly, the model of the FPS game acoustic ecology may now be modified to include the concepts discussed in this chapter as shown in *Figure 4.1*.

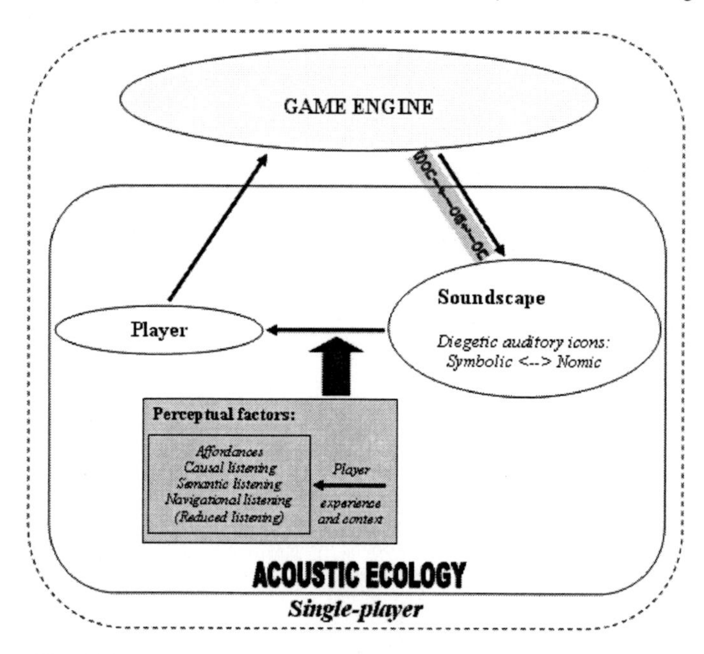

Figure 4.1. The FPS game acoustic ecology model expanded to include aspects discussed in chapter 4 (single player).

The triangular, cyclic relationship between player, game engine and soundscape is now partly explained by the game engine triggering audio samples through a process of sonification and these auditory icons form part of the soundscape which the player hears. The meaning the player derives from such sounds is modified by a range of perceptual factors including functions of sound which, when combined with player experience and context, affect the player's prioritization of or use of affordances and modes of listening. What the model does not show is any contextualization with regard to images, game events, player input or spaces. Neither does it explain any of the possible relationships between players as may be explained by sonification nor how players are able to be immersed within and participate in the FPS game acoustic ecology. Such contextualization and any possible relationships modify the meaning of sound further and so the following three chapters are used to explore these factors. They build upon the ideas discussed here and, in so doing, continue the work of constructing the conceptual framework of the FPS game acoustic ecology prior to the exposition of the hypothesis in chapter 9.

Chapter 5

Sound, Image and Event

Introduction

Chapter 4 discussed a range of theories relating to how humans design meaning in sound and how they derive meaning from sound. Although it did touch on other areas as a preparatory exercise for this and subsequent chapters, generally the chapter dealt with meaning in the context of sound alone. However, sight-capable humans rarely operate in any environment, including that of FPS games, solely by sound. This chapter expands upon that context of sound *sui generis* by examining some of the relationships between sound and image and sound and causal event or action; that is, how meaning may be refined or developed from any that the sound alone may have to a meaning that includes information about images and events associated with the sound. At the end of this chapter, the model of the FPS game acoustic ecology is further developed to account for sound in the FPS game in the context of image and game event.

The relationships between sound and image and sound and event are important for the hypothesis despite the hypothesis, at first sight, being concerned solely with the acoustic ecology of the FPS game. Sound in the game is designed, in most cases to work in conjunction with image and or event. Reverberation and localization of sound (discussed in chapter 6) work cross-modally with the visual level design to provide confirmation of the apparent scale, volume and materials shown on the 2-dimensional screen. Using Laurel's (1993) term, sound and image should have equivalent *resolution* (pp.164—165). Some types of sound work with image to provide a sense of location and temporal period (see chapter 6). Player expectation and audiation (see chapter 4) are based upon the assumption that certain visual objects in the game world should sound in certain ways and that events or actions in the game should have a sonic consequence consistent with the action depicted. Finally, sound works with image and event to bolster a system of perceptual sureties and surprises and these are key factors in creating a perception of immersion in FPS games (I talk in detail about immersion in chapter 7).

Clearly, the design of a relationship between sound and image and sound and event in any FPS game is of key importance during the development of the game, during the playing of the game and for an understanding of the game's acoustic ecology.[1] Unless the designers are striving for absurd or comical effect, there should be a consistent and expected mapping between game image and sound and game event and sound for each to reinforce the effect the other provides. If this is achieved, then, all other things being equal, the player is more likely to efficiently learn the parameters of the game in order to become immersed in the game world and to be able to participate in the game's acoustic ecology. In many cases, this is a matter of player control over sound production.[2] When I fire a weapon, I expect that the sound

---

[1] Nevertheless, it should be borne in mind throughout this chapter that not all sounds in the acoustic ecology will have a visible source.

[2] Production of sound, here means that the player is able to trigger audio samples from the palette made

produced not only sounds when I dictate but also matches what is to be expected from that weapon.[3] When my projectile hits another character, I expect the consequent sounds of pain, for those sounds to have differences depending upon the character being hit and for such sounds to confirm that my aim, as always, is true. Furthermore, a bullet ricochet off what looks like metal should have a different sound to a ricochet off what is a different virtual material within the game world.

This chapter, therefore, examines a variety of different disciplines pertaining to the relationships between sound and image and sound and event and extracts and develops concepts relevant to the hypothesis. Among the areas discussed are cross-modality, causality, the dissociation of sound from cause, acousmatic sound, synchresis and the indexicality of sound in the context of image. Many of the ideas under discussion have been heavily theorized in film sound theory and so frequent comparisons are made between sound in cinema and sound in digital games. Some of the relationships discussed here, and the taxonomies derived, are developed further in later chapters especially the relationships between sound and the depicted spaces and locations in the game world (chapter 6) and the role of sound in accomplishing player immersion in the FPS game acoustic ecology (chapter 7).

5.1 The relationship between the visual and auditory modalities

The visual and auditory systems are often treated separately because they are believed to be separate and distinct systems. Synaesthesiasts would dispute this and there is other evidence to suggest that, perceptually, the two systems have some relationship to each other. This interrelationship usually manifests itself in terms of perceptual localization and spatially shifts the perceived source of the heard sound to some point influenced by the visual system. In an attempt to evaluate the roles played by both auditory and visual systems in sound localization using an objective, quantitative approach, Witkin, Wapner and Leventhal (1952) found that "[t]he perceived direction of a sound coming from a visible source is determined through the use of both visual and auditory cues" (p.58).[4] Disregarding audio-only games, synergies between the auditory and visual modalities clearly have importance for the design and playing of FPS games and for the relationship between the FPS acoustic ecology and the visual game world and this is what this section begins to investigate.

Warren, Welch and McCarthy (1982) go further in suggesting that the two perceptual systems, visual and auditory, may in fact be one system with humans perceiving sound synchronized with image as one perceptual event. Notwithstanding this, cross-modality is the term used by the authors discussed below to describe relationships between the different sensory and perceptual systems and cross-modal confirmation is the situation where one sensory/perceptual system serves to confirm the sensations and resulting perceptions gathered by another sensory/perceptual

---

available by the game engine.

[3] Whether that expectation is founded on real-world experience or on prior knowledge of the game's conventions (as discussed in chapter 4).

[4] Interestingly, the researchers also discovered that "differences were observed between men and women, with women showing a tendency to locate the sound closer to where they saw it originate from than did men" (Witkin *et al.*, 1952, p.65) — the inference being that females depend more on visual cues for the localization of sound than do males. The article does not go further in suggesting whether the reason for this is biological, cultural or something else. This is perhaps an interesting point of departure for future research.

system.[5] The phenomenon of cross-modality, and the related system of cross-modal confirmation, between visual and auditory systems (modalities) is worth investigation as these two sensory/perceptual modalities are the ones most utilized when playing FPS games.[6]

In terms of the perception of films, Anderson (1996) states (arguing from an information model of cognitive theory as opposed to other models such as autopoiesis which argue for autonomy of cognitive function): "Perception is an information-gathering activity. And when it occurs in two or more sensory modes simultaneously, it is a process of information comparison, an active search for cross-modal confirmation" (p.82). This can work not just on the perceptual level, our eyes confirming by association what we hear or our ears confirming what we see, but also works at the higher cognitive level by making sense of a film's narrative (the sounds need not be diegetic or causally related to objects on screen). Anderson argues for this by giving an example of the discomfort felt when watching a silent movie with no musical accompaniment compared to the situation when an appropriate accompaniment is provided. The same expectant response (that there be sound in conjunction with image) may also be assumed to be true for modern digital games which, in most cases, are audiovisual rather than, as in the early days of digital games, purely visual.

Sound travels through air with a speed of approximately $344ms^{-1}$ (disregarding the effects of a range of factors such as height above sea-level, humidity and temperature, for example) while light has a speed of approximately $299,792,458ms^{-1}$. However, for sound and image that are delivered near simultaneously to the subject (by artificially delaying the visual signal), experiments have shown that sound is processed in the brain 40—60msecs. faster than image (Bussemakers & de Haan, 1998). Fitch and Kramer (1994) suggest that this is because the auditory system processes sound events in parallel whereas the visual system processes visual objects in series (pp.322—324).

This has two possible implications. Firstly, and as an example of cross-modal confirmation, it seems apparent that the visual system serves to confirm what the auditory system senses and perceives, which appears to be confirmed by the evidence of synchresis where one sound among the many heard is perceptually matched to the correct image on screen (for a further discussion of synchresis, see section 5.4 below). Secondly, it may be that humans prioritize sound sensations over visual sensations. If this is true, being, perhaps, a result of dangerous and opportunity-filled environments during human evolution when an omnidirectional sense would have advantages over other senses, then there are implications for the design and use of sound in FPS games where the player is in a similar, though virtual, 'hunt and be hunted' environment and where not all dangerous or

[5] Modality, in this sense, is not be confused with the use of the term in other areas such as descriptions of and relations between textual and narrative modalities in digital games by authors such as Burn and Schott (2004) or where modes are elements of a media text, its 'truth-claims' by which that text may be analyzed (Burn & Parker, 2003). In the context of this thesis, it refers to one of the five senses and cross-modality pertains to the relations between two or more of these senses — in particular, the relationship between sight and hearing.

[6] There is not yet any FPS game capable of providing the sensations of taste or smell and the sense of touch, as provided by crude haptic game interfaces, is little used in FPS games. I did once make the half-serious suggestion to some university students several years ago that it might be fun to couple FPS character pain and death with electric shocks delivered to the player but I don't think any of them took it seriously enough to implement this amusing idea in their subsequent game industry careers.

opportunistic events are available to be sensed by the visual system.

Laurel notes that cross-modality between the visual and auditory modalities, in conjunction with good kinaesthetic design, "is the key to the sense of immersion that is created by many computer games, simulations and virtual-reality systems" (p.161). She further notes another perceptual relationship between the two systems, one that likewise impacts on the sense of immersion within a game: "[W]e tend to expect that the modalities involved in a representation will have roughly the same "resolution" [...] A computer game that incorporates breathtakingly high-resolution, high-speed animation but produces only little beeps seems brain-damaged" (Laurel, 1993, pp.164—165). Such absurd juxtapositions can also be the gateway to humour — witness the use of sound in Warner Bros. cartoons for example or the display of iconic sound in Jacque Tati's films — but have little relevance in the serious intent of FPS run and gun games.[7] Additionally, such juxtapositions are not likely to lend themselves well to fostering a sense of immersion in the FPS game due to a lack of consistency between what is seen and what is heard (as I discuss in chapter 7).

More work needs to be done on the relevance and application of the cross-modality of the visual and auditory systems to virtual environments such as those found in FPS games. As Gröhn *et al.* (2001) point out: "The cross-modal perception of auditory and visual stimuli is explored mostly with animals [...] little research have [*sic*] been done in the area of cognitive aspects of simultaneous visual and auditory stimuli in dynamic environments" (p.15). Although this work discusses the topic in a theoretical manner, it is hoped that future work will develop empirical methods to explore it further because, *prima facie*, cross-modality is an important aspect of FPS games particularly for its potential effect upon immersion.

5.2 Causality

It may be presumed that sound designers will create a sound, and that it will be made available to the player or triggered by the appropriate game code, for a specific purpose. Broadly speaking, the intent is that the player accepts and responds to the main affordance offered by the sound in a manner that confirms its original purpose. An example, in the FPS game, is the sound of footsteps. The intent is that this indicates the presence of a moving character in the game world (whether the player's own character or another) and the awareness of this may be said to be the primary affordance offered by such a sound. There are other possibilities though; the sound may be that of fingers slowly tapping on a surface or it may, if particularly rhythmic, be the tick-tock of a clock. Chapter 4 mentioned that affordances are subject to prior experience and socio-cultural background. In order to recognize a gunshot in an FPS game as such, the player must have previously heard that sound in the correct context so that the link is made (and then remembered) between the sound and the weapon. This first hearing may be from personal experience, from other media such as the cinema or it may have been during the first playing of the FPS game. I have also previously stated that use of the causal listening mode is important in the context of the FPS game world. Combining the two points then, causal listening works with sound memory to assess the veridicality of the heard sound. In the

---

[7] FPS games designed for comic effect do exist but are very much the exception. Some user-designed levels for *Quake III Arena* (id Software, 1999) in which not only the visual design and characters but also the sounds, do exist mimicking cartoon-like platform games of the Nintendo and Sega type.

absence of image or of any other cues, the sound that is unrecognizable[8] is the sound which has not been heard before; the sound to which veridicality may be ascribed is that which has been heard, and categorized, before. It is veridicality of sound which helps create a believable and consistent acoustic ecology in which the player is immersed and may participate in.

This means that sound must have causality, must be as faithful as possible to its sound source, containing and retaining, from recording or synthesis through to playback, all the information required for the player to accurately perceive the cause and, therefore, the significance, of the sound. As noted by Gaver (Gaver, 1993a, 1993b) and other authors (Hahn, Fouad, Gritz, & Lee, 1998, for example), sounds are shaped by the properties of their source objects and actions and humans are able to assess these properties through sound. The sound should also contain information that enables the player to make accurate assessments of the environment (in so far as is possible) in which both he and the sound source are placed. I later discuss other factors that contribute to veridicality (section 5.2) such as synchresis (section 5.4) and localization of sound (see chapter 6), but here I wish to expand upon causality and to focus upon the ecological causality of sound that, through its sonic structure, informs the player about the game's acoustic ecology and the cause of the sound. It is possible, with the right modeling, to work in reverse and shape the characteristics of virtual objects, their motions and the sonic signatures of the surrounding game environment from the characteristics of sound thereby providing the 'coherent perceptual experience' that is fundamental to a perception of immersion in FPS games.

Reminding us that the scientific study of sound springs from research into musical acoustics (hence research that focusses often on frequency, intensity and time), Gaver (1993a) writes that little is known about other modes of auditory perception and that the traditional approach argues that knowledge about a sound event relies on memory and experience (pp.3—4). As he states:

> [M]aterial per se does not exist as for mechanical physics, but instead is separated into many other dimensions such as density, elasticity, and homogeneity. Nonetheless, people do seem to hear the material of a struck object, rather than these other properties (Gaver, 1993b, p.310).

In other words, where acousticians refer to properties of sound other than frequency, intensity and time, they do not reference the material of the sound source preferring to discuss this in terms of the general physical properties of the object such as mass, density, springiness and so forth. Gaver is right in stating that other people often describe a broad range of sounds they hear by the material of the sound source or by its action. Thus a sound may be a metallic clang, a wooden thump, a scraping action or a glass-like shattering.

Gaver describes a set of experiments in which subjects derived sound sources with a high degree of accuracy (up to 99 percent in some cases) from exposure to the sound alone and, when subjects were asked to describe what they heard, in almost all cases they described the sounds heard by using the materials or actions of the source (pp.18—20). It is important to note that, while materials and dimensions of

---

[8] By extension, and because humans normally attend to the material and causal properties of the sound, what is unrecognizable is more importantly the cause of the sound.

the sound source object may be encoded in sound (for example, larger objects tend to have lower frequency than smaller objects), the actions of that object are, for the most part, encoded temporally. That is, actions such as bouncing and breaking of objects are characterized by a temporal dimension which may be manifested by a pattern of discrete groups of sounds attenuating in series, such as for a bottle smashing on a hard surface (Gaver, 1993b, p.305). Gaver thus argues that information regarding some of the properties of a sound source is encoded in its sound and so, therefore, it is possible to design these properties into sound.

Modeling sound by material properties and dimensions is the basis for acoustic or physical modeling synthesis such as demonstrated in the Yamaha VL1 of 1994. With this synthesizer, it was possible to construct a sound where the strings of a violin are set in motion by a trumpet mouthpiece and amplified by the body of a flute in addition to other chimaerical sounds. Gaver goes further in suggesting that not only the materials of sound source objects can be modeled but also their dimensions and actions and he describes a set of algorithms to synthesize sounds based on all three of these. For example, he provides algorithms to model the salient sound features of struck bars, bars scraped over textured surfaces, objects falling into liquids and bouncing, breaking and spilling sounds.

Although FPS games employ almost exclusively audio samples rather than synthesizing sounds during gameplay, it would be interesting to speculate on the possibilities of synthesizing all sound according to Gaver's principles were the processing power available for it (as it no doubt will be one day). Boyd (2003) suggests that film sound designers are able to design and record a sound for every event or object in the film; they do not suffer from the same memory constraints that game sound designers labour under. There is, it seems, a tension that currently exists in the design of FPS games between wanting to employ unique sounds in a digital game and not having the computer memory or space on the game distribution medium to do so. This problem of memory and storage space (both RAM and ROM — for a discussion of this, see chapter 3) may be overcome, assuming the processing power is available, by the use of synthesis as described above. For example, footsteps on a variety of hard surfaces can be synthesized for *each* footstep from a basic model derived according to Gaver's algorithms with some minor adjustment of parameters each time it sounds to account for variables such as a different weight applied to each step, the speed of the walking or running action or the game environment.[9]

Breinbjerg (2005), writing about sound in *Half-Life 2* (Valve Software, 2004), suggests that, although the player is not able to confirm precisely the cause of a sound or the qualities of a sound source other than its rough features, sound in the game does inform about the source's salient features. For example, the sound of another's footsteps informs the player that a character or bot is in the vicinity but does not provide information about the weapon which that character is carrying or whether the character is friend or foe (context will help in the latter[10]). This would

[9] Some features of a synthesized sound may be imperceptible to humans and so not all aspects of the sound need be encoded — similar in principle to the 'lossless' encoding of higher-order mp3 algorithms which reduce audio data by discarding the imperceptible or least perceptible parts of the original music recording.

[10] In the case of the context being a single-player FPS game, the probability is that they will be enemy footsteps and, with experience, further information may be available as to, for example, the weaponry being carried. Although it is many years since its release, anyone who has played it, may still recall the massive stomping sound and the shiver of anticipation coupled with a tensing of fingers around the controller as the final, immense demon, armed with fire-

seem to support Gaver's contention that relatively simple algorithms modeling the salient characteristics of the sound event can provide an appropriate level of veridicality — veridicality in so far as the player, in the first instance, need only know that another character is in the vicinity before proceeding to assess the threat posed by that character.

Discussing veridicality (according to him, the central organizing principle of the visual system), Anderson states: "[A]n individual's perception of the world needed to be a very close approximation of that world. It had to be accurate enough to act upon because the consequences of error were severe" (p.14). If it may be surmised along similar lines that veridicality is the central organizing system of the auditory system as well, then a deliberately reduced sonic veridicality may be all that is required to place the FPS player in a heightened state of alertness. (This complements my previous suggestions regarding synthesis and the imperceptibility of certain parameters of sound — the synthesis may be simplified while still retaining an acceptable level of veridicality.) This perceptual veridicality is, of course, helped by the fact that the (experienced) player expects to hear footsteps or similar sounds in the FPS game and is therefore already in Bruner's (1957) state of perceptual readiness (pp.135—137).

The ability to assess the causal properties of the sound or the dimensions of the sound source is usually a process of categorization. Upon hearing the striking of a metal bar with a hard object, a person does not describe it in terms of amplitude or frequency; it is described in terms of the properties of the sound source and other parameters such as its location — *it has a hard, sharp metallic ringing, is nearby and is coming from that direction.*[11] The clarification of the question *what's that sound?* is never *what, the one at 105dBA with a fundamental frequency of 565Hz and a large proportion of high frequency energy that's 10m away at an angle of 47° from the meridian?* but is more likely to be *what, that loud metallic clanging coming from that room over there?* If, upon examination, the sound source in question turns out to be a bell struck with a mallet (and the final response to the question posed above is that it *is* a bell), then the sound is predictively veridical. As Bruner states:

> By predictive veridicality I mean simply that perceptual categorization of an object or event permits one to "go beyond" the properties of the object or event perceived to a prediction of other properties of the object not yet tested. The more adequate the category systems constructed for coding environmental events in this way, the greater the predictive veridicality that results (p.129).[12]

Ancient peoples have long possessed and exploited the ability to dissociate sound from its causal event — the design of Romanesque and Gothic cathedrals for example or the use of natural echo chambers such as caves — and this may be viewed as an instance of acousmatic sound in its Pythagorean sense as I describe in section 5.3. The mysteriousness of such causally dissociated sounds is exploited for

---

balls, hunted them down in the last level of *Doom* (id Software, 1993).

11 Useful not only for causal listening but also for navigational listening (see chapter 4).

12 This trope may be sequenced further. The prediction that the metallic ringing is a bell may lead to the prediction that there is a church with a tower which may lead to the assumption that there is a human settlement nearby and so on. See chapter 6 for further discussions on this in the context of soundscapes.

different purposes in horror films where the fear which may be experienced by the audience in large part derives from their inability to causally determine just what that horrible scuffling noise is that sounds in the dark.[13] These atmospheric sounds are also used in game design, for terror or for more general symbolic purposes, and, in the context of natural ecologies, have been described by Schafer (1994) as archetypal sounds, sounds possessed of a symbolism, inherited from ancient times and that have a mystery about them (pp.9—10). I myself, have used such sounds in the design of a *Quake III Arena* custom level *Grim Shores 3: Atlantis* (Grimshaw, 2001) in which I provided causally indeterminable whisperings and chantings wrapped in a long cathedral-like reverberation matched by the gloomy catacombs and pseudo-pharaonic interiors of the level.

For a believable acoustic ecology, the ecology must be comprised, in the main, of veridical sounds and, to a large extent, veridicality is a matter of causality; the ability to assess the material properties of the object causing the sound or the nature of the source action. Although certain causally indeterminable sounds may contribute to feelings of horror or otherwise aid in the creation of atmosphere, most sounds in the FPS game are causal. Believability is important to a perception of immersion (and thus the variety of spaces in which the player is immersed — see chapter 6) and the ability to participate in the acoustic ecology as I discuss further in chapter 7.

Almost all modern FPS game sounds of the diegetic variety are designed or selected for on the basis of causality — that is, the intention is that the player, upon hearing the sound, will be able to ascertain its cause whether that cause is a source object or an action. Causality in the FPS game is different to causality in the real world. In the latter (where no recording or amplification has occurred), the sound sensation (as opposed to sound perception) that is heard is derived direct from the object or action with no intervening and mediating factors other than artefacts such as reflections off other objects and surfaces adding further elements of veridicality to the sound. In the FPS game, many other factors come into play in an effort to convince the player that the heard sound emanates from a group of mute pixels on the screen. Although the real cause of the FPS game sound is the operation of computer code (which itself may be triggered by player input), the player is asked to believe that the cause is actually an object or action within the game world, a belief that is similar to Chion's (1994) *audio-visual contract* in cinema.

With reference to causality, diegetic sounds have two classes — sounds that are noncausal and sounds that are causal. A noncausal sound bears no physical resemblance to the source object or action and, if described as an auditory icon, is therefore usually a symbolic or metaphorical auditory icon (see chapter 4). Having no physical resemblance or relationship to the source object or action, the sound cannot be used to predict the properties or qualities of that source object or action. These sounds, in the FPS game, tend to be in the minority and include, for example, game status notification sounds such as the flag status musical chords in *Quake III Arena*. With experience, a player will learn what the source object or action associated with this type of sound is, hence the causality of the sound in a virtual, computer code-driven sense, but these sounds are not strictly causal as they cannot be used to assess the properties and qualities of the source object or action — in

---

[13] A fear heightened further by the director's (and not the audience's) control over the visualization or not of that sound. In fact, this very inability on the part of the audience to control the visualization of sound may be another factor in the perception of fear.

reality, a dropped flag does not emit a musical chord.[14]

A causal sound is one that fulfills the requirements of causality and so is a sound that bears some physical relationship or resemblance to the source object and which, therefore, can be used to judge the properties and qualities of the source object or action. As with any causal sound in the FPS game, this physical resemblance or relationship is an illusory one as the causal sound has only a contractual indexicality to the source object or action within the game world and not a real indexicality as would be the case for a causal sound in the real world. Nevertheless, if we abide by this contract and accept the illusion offered us, then a causal FPS game sound can be defined as a sound that bears some illusory physical resemblance or relationship to the objects and actions presented within the game and, furthermore, can be used to predict some of the properties and qualities of those objects and actions.

In *Quake III Arena*, there are a variety of surface materials within the game world that generally fall into the categories of solid or liquid. Whenever a character moves from walking or running on a solid surface (usually metallic or stone within the game environment) into water, the usual pitter-patter of virtual feet is replaced by sounds of splashing. This is an example of an action's sound being dynamically causal in changing its sound to match the immediate vicinity. Some FPS game engines (in conjunction with the appropriate audio hardware and software, such as A3D from the now defunct Aureal) allow for a similar dynamic causality with respect to reverberation; in this case, sounds such as footsteps and gunshots will have real-time reverberation and reflections applied dependent upon the dimensions and materials of the game space around the character. Other game engines (or game engines combined with less capable audio hardware) only have a static causality (in so far as reverberation is concerned) in which the sound designer must supply a pre-reverberated sound as an indicator of the acoustic properties of the game locale.

Causal sounds are part of the system of perceptual cues (or sureties) and surprises and, therefore, part of McMahan's (2003) system of *perceptual realism* which is an important factor in creating a sense of immersion within the game world (pp.75—76) and which is discussed further in chapter 7. Sounds that function as perceptual cues are causal because they are expected details in the game environment and thus must match the objects and actions depicted in that environment. Sounds that are perceptual surprises may be causal or may be something else which, once the player has fathomed the source of a perceptual connector for example, may become causal through the player's experience. They are not noncausal as they may bear physical resemblance to the source object or action. As an example, querying what that repetitive, mechanical noise is that can be heard in the distance, the player discovers it emanates from a room full of machinery.

Sounds that are initially noncausal may, with experience and in the player's perception of that sound, become causal, hence the preferred term *quasi-causal sounds* forming part of a continuum with noncausal and causal sounds and thus apportioning degrees of causality. A quasi-causal sound hints at its source object or action, inviting confirmation on the part of the player of the cues it provides. The experience gained in doing this converts this sound into a causal sound for that player (although it may remain quasi-causal for other less-experienced players). The

[14] Hence, as discussed in chapter 4, this audio sample is a symbolic auditory icon.

concept of quasi-causality and the ability for such a sound to mutate into a causal sound support the notion that the meaning of many sounds in the FPS game is derived through a learning process as the player engages with the acoustic ecology (see chapter 4). However, not all noncausal sounds can become quasi-causal but are destined to remain mysterious, symbolic or archetypal sounds and may be used to add a miasma of unsettled atmosphere to the game (see chapters 7 and 8).

This section has demonstrated that sound can be assessed and engaged with in terms of its causal properties. Indeed, the normal mode of listening in an ecology is causal listening (which may be combined with other modes such as navigational listening). It is important, therefore, that most sounds in the FPS game acoustic ecology be causal; the player should be able to derive the causes of most sounds in order to utilize the affordances offered by such sound. In this prioritization of the causal listening mode in the context of the FPS game, it is entirely appropriate to stress, as Stockburger (2003) does, that the reduced listening mode (attending to the frequency of the sound and amplitude, for example) is little used and that causal listening and, to a lesser extent, semantic listening modes have more utility.

5.3 Acousmatic/visualized sound

This section investigates the concepts of acousmatic and visualized sounds and the possible application of such concepts to the FPS game acoustic ecology. Although initially formulated for electro-acoustic composition and developed in film sound theory, theorists, such as Stockburger, have also made use of them in order to analyze digital game sound. The media are very different though. Particularly, it is the case that the game player has far greater control over the sounding of many sounds than either the listener or the spectator have in the former two media. In the FPS game especially, the player, in many cases, is able to turn towards, and thus display, the cause of a sound heard off-screen.

Acousmatic is a Pythagorean term that describes the distance between the point of hearing and the point of origin of sound, specifically the distance separating disciples from an intoning priest hidden behind a curtain (Pythagoras himself apparently used this technique when lecturing to students to force them to concentrate on his words). It was taken up by the electro-acoustic and *musique concrète* tradition in an attempt to remove all possibility of causal listening and force a concentration (using the reduced listening mode[15]) on the qualities of the sound object itself (Dhomont, 2004). In some film theory literature, acousmatic sound is termed off-screen sound, but this is a problematic term for other theorists (Metz, 1985, for example) while Chion comments that technological developments in cinema sound, such as surround sound, blur the distinctions between on-screen and off-screen sound (pp.129—131).

For film sound theorists such as Chion, an acousmatic sound is a sound for which there exists no visually identifiable source on screen and an acousmatic situation "intensifies causal listening in taking away the aid of sight" (p.32) despite the hopes of electro-acoustic composers that acousmatic situations would intensify reduced listening. Various terms have been proposed for its opposite (such as 'direct') but Chion's term 'visualized sound' has the advantage of combining the twin worlds of

---

[15] There is an irony here in the use of the term acousmatic in the context of film and, especially, digital games where, as previously stated, causal listening is prioritized at the expense of reduced listening.

image and sound (p.72). Nondiegetic sound, according to Chion, is acousmatic sound but with no connection to the storyworld inhabited by the film (pp.71—73).

In cinema, the audience is sonically passive in the sense that it has no control over the visualization or de-acousmatization of sound. Such decisions are left to the director (certainly in terms of direct sensation of sound from the film — perception and cognition of film sound is certainly subject to external factors both private and public) and, once in the final film print, cannot be changed.[16] Indeed such directorial decisions have a direct influence on the film script, scenography and, therefore, final form of the film and, as Curtiss (1992) has pointed out, in some cases have a more fundamental influence on the genesis of the film.

Here, perhaps, are two important differences between film sound and game sound. Firstly, the game player controls the visualization or de-acousmatization (and re-acousmatization) of many sounds by adjusting his game position (by virtually turning the head or body within the game environment). This is what Stockburger terms kinaesthetic control of sound objects (p.9). Secondly, and what has not been discussed by others as a point of difference between film sound and game sound, is that, to all extents and purposes, there is an almost infinite number of combinations of acousmatic and visualized sounds in varying proportions. No two playings of a game will produce the same soundtrack[17] — in film there is only one soundtrack with the authoritarian stamp of the director. Without access to the game code or ability to design his own levels, there is no suggestion that the player can choose or modify the palette of sounds available or that all sounds are subject to the player's control.[18] What is being suggested here, though, is that the player can have a great deal of control over the sounding of many sounds within the game. Furthermore, a multiplayer FPS game produces as many soundtracks as there are players, each customized, as it were, to the player's particular location in the game, relationship to other players, role in the game and personal audio settings. The audience in a cinema is delivered the same soundtrack at the same time (with allowances made for differing perception of the sound due to factors such as audience noise, relative position to loudspeakers, for example).

The player, with kinaesthetic control, is an active participant in the drama being played out on screen. In effect, the player is a performer and 'all the game's a stage' (see chapter 9 for a wider discussion on the FPS game acoustic ecology as theatre). Digital games differ in many respects to other audiovisual forms but perhaps the single most important difference (especially with regard to cinema) is the level of interactivity. The act of playing the game requires physical input from the player and, in an FPS game, this act usually results in a sonic consequence that contributes not only to the sounds that player hears but also to the sounds, in a multiplayer game, which other players hear (thereby potentially affecting their responses). This is a vital point to consider in light of the hypothesis suggesting that the player is an integral and contributing component of the FPS acoustic ecology and is returned to in chapter 7.

Sound in the FPS game, therefore, may be acousmatic sound or visualized sound

16 Bar editing the film print itself or through censorship or dubbing.

17 This, implicitly, is the basis for the term and suggested practice of 'sonichima' (http://www.selectparks.net/modules.php?name=News&new_topic=29).

18 On a desktop computer, it is possible to replace game supplied sounds in the installation directory.

and, in many cases, an acousmatic sound may become a visualized sound as a result of a deliberate act on the part of the player. This is an important distinction to make between sound in the FPS game and sound in cinema. Although the terms acousmatic sound and visualized sound are widely used in cinema, they are valid terms to use in FPS games with the understanding that, in most cases, it is the player and not the director who is in control of any change in state between the one and the other. Because this control takes place during gameplay (the FPS game soundtrack is not a fixed entity as is the soundtrack in cinema), this has implications for the interactivity of the game. The FPS game acoustic ecology is founded on the basis of sonic relationships not only between players but also between players and the game engine. The ability to acousmatize and visualize sound in the game world is an important instance of this latter relationship because it is one demonstration of the player using the game engine in real-time to transform the acoustic ecology of which he is a component.

5.4 Synchresis and point of audition

One similarity (with some points of difference) between film sound and game sound revolves around the idea of synchresis. Synchresis is the melding of synchronous aural and visual objects such that we believe them to be one and this phenomenon is not fully automatic but is a product of meaning — context, volume, rhythm all play their part (Chion, 1994, pp.63—64). Anderson's synchrony is identical and is described by him thusly: "[I]f the auditory and visual events occur at the same time, the sound and image are perceived as one event" (p.83). Synchresis is similar to the audio-visual proximity effect as used in acoustics and related disciplines which is explained by Czyzewski *et al.* (2001) as: "In most cases the video "attracts" the attention of the listener and, as a consequence, he or she localizes the sound closer to the screen center" (p.546). Psychologists would term this the ventriloquism effect and, to some extent, it can be explained in terms of cross-modal confirmation as discussed above.

In the real, physical world, if an object makes a sound, the sound almost always originates from that object (that is, *direct* sound — to simplify matters the added probability of reflected sound is ignored here) and the physical sensation of that sound (assuming no technological intervention) is perceived to emanate from that source object. The accuracy of this perception within the auditory field of the listener is variable, being quite precise (to within c.3°) directly in front of the listener and in the horizontal plane and becoming less precise as horizontal, vertical and lateral listening orientation changes. This perceptual accuracy is intensified by the practice of nutating — minor movements of the head to take advantage of differences in signal path length and signal intensity as the sound arrives at our stereo auditory system — and is aided further if the source object is mobile (the eyes, in this case, guiding the ears).

In cinema, sound very rarely originates from the location of the source object on screen — this occurs to some degree of accuracy only in the most sophisticated (and therefore, rare) multichannel loudspeaker systems in which some loudspeakers may be positioned behind the screen. What is more usually the case is that sound, apparently emanating from an object on the screen, actually originates from one or both of a set of stereo loudspeakers positioned in front of the audience (or from more loudspeakers around the audience depending on the sound system used by the cinema).[19] Providing the film sound engineer has made a good enough job of matching the sound to the context and has taken care to synchronize sound to image (or, as undertaken by the film editor, the reverse for much of the assemblage of music video), then the film audience accepts the illusion that the sound event is mapped onto the visual event becoming one event (the sound does not necessarily have to have been derived originally from that object but should match the audience's expectations of what such a screen object should sound like). This is synchresis.

The situation in FPS games is much the same. Physically and technically the standard visual and auditory interfaces are not that dissimilar for games and films. A 2-dimensional screen of asymmetrical aspect ratio that occupies a portion of the field of view and a set of stereo transducers that are either loudspeakers or are headphones (there are refinements of this, particularly with the auditory interface, but this is the standard film and gaming setup). If the context makes sense or sound is synchronized to motion on screen, then the player experiences the effect of synchresis — a digital, audiovisual game could not operate otherwise.

There are some differences in the validity of synchresis between the two media, the first two being only potential depending on the game player's auditory interface. Firstly, if the game player is using headphones, conscious or unconscious nutating will make no difference. Of course, the same would be true if a film spectator were wearing headphones but this is far more likely to be the case for a game player (the manual for the game *Myst* (Cyan, 1993) makes the explicit recommendation that the player wear headphones).

Secondly, recommendations aside, many FPS players prefer the use of headphones to stereo or surround sound systems because localization and, to a lesser extent, depth perception are more acute while sound from the game environment is enhanced at the expense of sound from the user environment (computer fan, barking dogs, smoke alarm for example).[20] Indeed Morris (2002), talking of a typical gamer's hardware setup and location, says they often *have* to use headphones to "create their own dedicated sonic space which, not surprisingly, increases the immersiveness of the gaming experience" (p.86). And Schafer rather poetically describes the headphone listener thus: "[H]e is no longer regarding events on the acoustic horizon; no longer is he surrounded by a sphere of moving elements. He *is* the sphere. He is the universe" (p.119). There is a phenomenon with headphone use known as in-head localization where listeners fail to externalize sound and it is apparent that adding environmental cues such as reverberation to dry sound aids in externalizing that sound so that the listener hears it as coming from outside the headphones rather than inside the head (see Wenzel, 1992, pp.83—87). It is likely to be the case that synchresis plays a role in externalizing sound heard over headphones where that sound is juxtaposed to moving images.

Of significance to FPS game sound are film sound theory ideas on points of audition (POA) and internal auditors. For Chion, because sound is omnidirectional, there is no single point of audition, rather a zone of audition (pp.89—92). In a film, techniques such as close-up or focus and reverberation, are usually a strong

---

19 In the case of television sets, most are monophonic with one loudspeaker positioned outside the screen.

20 This sensory immersion is discussed further in chapter 7.

indicator of which on-screen character a sound is intended for but, for the FPS game, the answer to Chion's question *Who, on screen, hears what I hear?* (pp.89—92) is *I hear what I hear* (and potentially other players in a multiplayer game). Because there is only ever one point of view (POV), there is only ever one POA.[21]

For those film theorists who do recognize a single point of audition in cinema, very broadly speaking, the notion of the auditor in talkies has historically developed from that of an impossible auditor to an external auditor to an internal auditor — a discussion of these points of audition may be found in a range of authors (such as Altman, 1992; Bernds, 1999; Doane, 1980; Lastra, 2000). The sound heard by an impossible auditor was amusingly, but accurately, described by John Cass in 1930 as "the sound which would be heard by a man with five or six very long ears, said ears extending in various directions" (quoted in Doane, 1980, p.54) and the resultant mismatch between sound level and the size of image as seen on screen came about through the spot placement of primitive microphones on set coupled with rudimentary balancing techniques. The later attempt to match film sound scale to film image scale (the external auditor) is analogous to how humans hear real-world events and was accomplished through the use of a single microphone placed close to the camera. However, this seemingly commonsense approach is in conflict with the requirement that the all-important film dialogue be intelligible no matter how far the camera is from the speaking characters or what other sounds are competing with it. And so the concept of the internal auditor was created in which the spectator vicariously inhabits the body of an on-screen character who will hear *for* them and this was achieved through the use of spot microphones (or, later, dubbing) wherever dialogue intelligibility was required.

However, modern FPS games are characterized (and, in fact, recognized as FPS games) by the presence of a pair of hands (or one hand) holding a weapon that is placed at the centre bottom of the screen receding perspectivally into it. These are always present and become an extension of the player into the game especially when they respond to the player's actions with shooting or reloading animations (and headphones aid in this propulsion into the game world by encasing the player's auditory system within sound from that game world). In (live action) film, not only are such prosthetic extensions rare (in practical terms alone there are difficulties in always having a pair of actor's arms crowding the front of the camera and cameraman) but, such prostheses, where they do appear on screen, never respond to the spectator's wishes.

Furthermore, a member of the film audience is always aware of his physical presence occupying a physical space with, usually, other bodies present some of which will be seen (heads blocking the screen or bodies on the periphery of vision) and most of whom will be heard (laughter, opening of packets of sweets). There is a lack of physical interactivity in film compared to games (especially when games are played with headphones as mentioned above) resulting in different modes of immersion such that the film spectator is always aware of and always torn between the existence of two worlds — that which they inhabit and that which they spectate. There may be critical involvement and some mental immersion in the film but there is no prosthetically corporeal projection into the film world in the sense that this is my world to do what I want with, to explore how I wish.

---

[21] POA is related to player contextualization in a resonating space and player immersion in that space. See chapter 6 for a discussion of spaces in the FPS game's acoustic ecology and chapter 7 for a discussion of player immersion.

In an FPS game, there is no sense of an external or internal auditor. This is due not only to the low level of prioritization given to dialogue in the FPS game (the exigencies of which, in film, led to a variety of artificial auditors as outlined above) but is also due to the first-person nature of the FPS game. The camera and microphone are united once more, embodied in the player who sees in first-person perspective and hears in first-person perspective, all from the same point of audiovision under the control of that player, and image/sound scale is natural (rather than distorted for the benefit of an internal auditor). Although this might suggest that this is a return to the sound conventions of the early talkies, this is not the case because there is no external auditor in FPS games. The player *is* the auditor, hearing neither from some location external to the gameplay nor through the mediation of other game characters but being immersed in the game world;[22] I hear therefore I am.[23] The auditor in the FPS game must therefore be considered as a *first-person auditor*. This is an extension, or a combination, of the notions of both external and internal auditors in cinema. Camera and microphone are co-located and, rather than an invisible character within the film's diegesis who hears for the spectator, the prosthetic limbs on the screen encourage the assumption of prosthetic ears on the player's character in the game world. Importantly though, and unlike any of the forms of auditor in cinema, this first-person auditor has kinaesthetic control and navigates at the command of the player.

To bring this back to synchresis, there are aspects of synchresis within an FPS game that directly relate to the first-person singular that is the player. *That person is shooting at me* (I can hear a shot and I can see a muzzle flash), *That tank is too far away for me to be concerned about* (a small screen image in relation to other screen objects and a quiet sound) or *I should get off the gravel onto the grass next to it before I'm heard by the enemy* (that crunching sound obviously comes from me running on that grey strip on the screen). The audio and visual feedback is so directed to the first-person that the game's visual monitor and, more especially, audio transducers become, in effect, a part of the human proprioceptive system. Headphones are substituted for our proprioceptive auditory system by blocking out sounds external to the game world, like our own breathing, such that the breathing of our character, our *self*, in the game is substituted instead. Here, the character's ears become aural prostheses equivalent to the visual prostheses shown on the screen. This hearing presence in the FPS game world has important implications for player immersion which will be discussed further in chapter 7.

Synchresis can be usefully summarized as the eyes guiding the ears and therefore has some validity for the analysis of sound in FPS games. However, synchresis relies on the fact that the source object is in view at all times. This is all very well for cinema where the director is in control or for source objects seen on the screen in FPS games but when a player can kinaesthetically control the de-acousmatization of sound originating from unseen objects within the game, synchresis plays no part until the source object is seen on screen. In this particular case, it might be said that the ears guide the eyes. Synchresis, then, is useful as an approach to understanding

---

22 As I point out in chapter 6, being *physically* immersed in one of the resonating spaces of the game's acoustic ecology.

23 Although Gröhn *et al.*, in the context of navigation in virtual reality environments put forward the interesting hypothesis that it may "be useful to separate viewing and listening points" (p.17), I am unaware of any FPS game offering vicarious points of audition in similar manner to the use of on-screen radar or schemata of the whole level displaying the position of teammates. Electronic eavesdropping equipment (or, indeed, the character role of spy) may be interesting additions to future FPS games.

some aspects of the FPS game acoustic ecology because, while sound may inform about the game world beyond what is depicted on screen, it also works together with the image on-screen in immersing the player within that world. In reality, sound from the game does not emanate from the pixels on screen but from external transducers and synchresis provides answers as to how players can perceptually centre heard sounds from the acoustic ecology onto seen images in the game world. Identifying the player as first-person auditor has significance for player immersion within the FPS game acoustic ecology.

5.5 The dissociation of sound from cause and the question of originality

This section explores synchresis further and reverses it by discussing what happens when sound is not perceptually associated with an image. It also explores the interplay between originality of sound and reproduction of sound — sound in the FPS game is indexical to player actions, there being no use for arguments concerning the originality and reproduction of sound in this context, and this strengthens the case for player immersion within and participation in the acoustic ecology. Ultimately, the player is encouraged to believe that the sounds he sounds in-game originate from his actions.

It has been noted by several authors (Ballas, 1994, p.80; van Leeuwen, 1999, pp.167—168) that one of the peculiar properties of sound, one that vision possesses to a lesser and less malleable extent, is that it can be dissociated from its causal origins or context — a phenomenon that Schafer terms *schizophonia* (pp.88—91). The physical dissociation of sound from its causal event is enabled by recording and telephony and so is a relatively new phenomenon.[24] In *Terminator 2* (Cameron, 1991), the shotgun sound is comprised mainly of the recording of two cannons (Palmer, 2002, p.9). Clearly, the sound designers of this film were looking for something that packed more of a physical punch than the relatively puny sound of a shotgun (something Palmer dryly refers to as Hollywood's "law of enlarged firearm calibre" (p.9)). Because the cannons' recording is synchronized with the image on screen and sounds gun-like, the sound is accepted as issuing from the shotgun image on screen and is a good demonstration of Chion's principle of synchresis — further examples may be found in LoBrutto's (1994) interviews with film sound designers. Although lacking indexicality, the sound tends to the indexical with increasing verisimilitude. In fact, the case could be made that the cannon recording now possesses an absurd indexical relationship to the shotgun. It can be argued that the experience, for many film spectators, of the sound of a shotgun comes solely from films and that years of Hollywood conditioning have led them to expect to hear that sound whenever a shotgun-wielding actor appears on screen. With no experience of a real shotgun sound, this *is* what a shotgun sounds like.

Furthermore, for synchresis to occur, the properties and characteristics of the sound used must not only be relatively close to the audience's expectation, their audiation, of what the object on screen should sound like (which itself assumes some familiarity with that object or class of objects on the part of the audience) but must also be supported by further visual cues pointing to the object in question. Thus, for example, the sound of a dog barking is not likely to be synchretized to the image of a

24 Certainly by electronic means.

shotgun on screen unless that shotgun is the only moving object on the screen and is moved or fired in time with the barking. Such deliberately absurd synchresis is widely used in films (both live action and animated) for comic effect as mentioned above but is rarely found in FPS games.[25]

Nevertheless, despite any physical or conceptual dissociation from its cause (and by extension its environment), sound still retains information referring to that cause and environment.[26] Where synchresis occurs on the screen, this information, if the sound is not too far removed from audience expectation, will either be conveniently mapped to the image on screen or will be ignored. In the absence of synchresis (or the absence of the image itself), that information is free to be interpreted and mined for meaning dependent upon training and experience. A recording of a dog barking (not juxtaposed to an image of a dog) is likely to be recognized as the sound of a dog barking the world over.[27] A recording of a dog barking in a large concert hall with its attendant reverberation may present problems in identification for those cultures having no aural experience of such cavernous and reflective spaces.

Here, we come up against the question that has vexed many a sound theorist whether that theorist works in the area of music recording or film sound recording: *Is a recording of the sound a copy of the original or a representation of the original?* The question arises out of Benjamin's (1936) oft-used dictum that "[t]he presence of the original is the prerequisite to the concept of authenticity [...] The whole sphere of authenticity is outside technical — and, of course, not only technical — reproducibility" and from a concern for the mediating effects of technology (recording and reproduction technologies) and space (the space the recording is played back in). Such mediation can be deliberate, as when 'tweaking' the equalization of the sound upon recording or reproduction, or not deliberate, such as artefacts introduced by the inherent distortion of electrical equipment or by the playback space. The issue is discussed in the work of many authors (such as Chion, 1994; Lastra, 1992; Warner, 2003), but here I wish to engage with one particular discussion which leads me to suggest that the diegetic sound, or at least many sounds, heard when playing an FPS game is always the original.

Some theorists argue convincingly that the notion of an original sound (in the presence of or contrasted to a recorded sound or reproduction) is theoretically worthless. Lastra takes Edison's view that the original of a sound recording is not meant to be heard by anyone as it is simply "one stage of a multistage representational process" (p.128). If one asks a person to describe in detail a sound heard previously, they will find the task impossible other than by making use of the broad brushes of generalization, simile and metaphor. Thus the sound may be colourfully described as *a high-pitched squawking sound like a parrot being strangled.* Those more versed in audio jargon and with better trained ears may be able to make a more accurate attempt but will still fall short of even a reasonably faithful description of the original sound unless they are an expert mimic.[28] Because sound exists solely in time and is ephemeral and dissipated once sounded, it is remembered only in outline and as caricature.

---

[25] A genre ripe for the use of such comically absurd juxtaposition is that of children's games.

[26] Assuming any audio processing is not sufficient to disguise the origin of the sound.

[27] With the possible exception of those African cultures whose only canine experience is with the Basenji breed.

[28] At which point it is no longer a description of sound but a sound in itself.

For Wurtzler (1992), a representation posits "an absent original event" (p.88). It matters not whether that event existed in reality (a recording of a concert for example) or whether the event occurs in fiction. Hence, as he explains, the audiophile's "fetishistic relationship to the means of representing [in order to gain] increased access to an original performance event" (Wurtzler, 1992, p.88). Théberge (1989) traces a transition, delineated by the introduction of the multitrack recorder, from a process of documenting (by recording) an original event to a process of construction of the event (enabled by overdubbing, editing and sampling for example): "[A] shift in recording aesthetics away from 'realistic' documentation of a musical event to the *creation* of one" (p.104). To this Wurtzler adds a third stage (brought about in part by the development of virtual reality systems which he describes as a representational technology), viz. a representation for which there is no original event: "[C]opies are produced for which no original exists" (p.88).

Unfortunately, Wurtzler does not supply the answer to the obvious question: *if there is no original, what are these copies of?* His discourse is increasingly obscure and ultimately self-defeating; if, as he says, a representation presumes an absent original, the model of virtual reality (to which the virtual environments of FPS games are closely related) is not, as he states, a representational technology if the original is neither there nor absent, but simply does not exist. There may be the argument that audio samples in an FPS game are representational (mediated by recording and reproduction) of an original sound event. But the argument cannot be applied to sound in digital games that is synthesized (uncommon in FPS games but common in games of the platform genre for example) or to the sound that is the sum of all sounds heard during gameplay because this is always a unique sound event, the original performance of the 'sonic composition' of the game, that is dependent upon the unique genesis of that particular gameplay. The way to resolve this crisis is to cut through the Gordian knot by stating that this sonic composition (which may be referred to as the soundscape[29] and which is examined further in chapter 6) *is* the original sound event and therefore the *authentic* sound event — 'authentic' as in Benjamin's use of the term. To prove the point, it cannot be replicated, because each playing of the game produces a different sonic composition, unless that soundscape is recorded and replayed at which point arguments over authenticity and representation versus the original may be applied.

If the causal event, the source, of any sound in the FPS game is deemed to be the cause or source of the original recording, then it may be stated that all sound in the FPS game is dissociated from its causal event. However, this statement ignores two important points about the medium that set it aside from the use of recorded sound as heard in music playback systems or in the cinema for example. Firstly, production[30] of much of the sound of the FPS game, particularly the interactable sounds (as opposed to environment sounds which are sounded by the game engine), is under the control of the player and, secondly (and this is a consequence of the first point), the production of these sounds is a real-time process. Most interactable sounds, (and some environment sounds — see chapter 3) are under the control of the player and thus may be said to be indexical with the player's actions — they require the player's actions in order to sound — and in this case, therefore, it may be more correctly stated that sound in the FPS game is rarely dissociated from causal

29 A term originally proposed by Schafer.

30 That is, the sounding of the audio sample as opposed to the creation or recording of the audio sample.

event.

This lack of dissociation of sound from cause in the FPS game is fundamental to the hypothesis. Of importance here is believability; a credible acoustic ecology paves the way for player immersion which itself paves the way for player engagement with and integration into the game world. Every (sound) effect must have a cause and this must be a credible cause within the accepted rules and conventions of the game. These are not written down in any manual but are received through experience in playing FPS games and through real-world experience (see chapter 4). The firing of a weapon must produce a sound and running on gravel should produce a different sound to running on grass, for example. What the specific sounds are that are used depends on several contextual factors, including the premise of the game, but the believability of these sounds is greatly dependent upon their relationships to the game's images. It would be perfectly acceptable to have a chicken squawking to replace the sound of an airplane engine as long as this absurdly comic juxtaposition is extended to other sounds and if the sound and context is consistent with each playing of the game — with experience, this becomes a part of the received rules and conventions for that game (although, as previously noted, comical FPS games are rare). Such juxtapositions become less comical and more believable if the sound matches the image (for example, the chicken squawk occurs whenever the player flies a chicken through the air) which brings me to questions of sound/image indexicality, iconicity and isomorphism; the subject of the next section.

5.6 Indexicality, iconicism and isomorphism of sound

This section discusses notions of the indexicality, iconicism and isomorphism of sound in the FPS game through the discussion of such concepts in film sound theory. Where it is suggested that sound in an animated film is never indexical, I suggest that sound in the FPS game, likewise visually an animated medium, is almost always indexical. This is because, where sound in animated films is predicated upon image, sound in the FPS game is predicated upon player action and thus is indexical to that action. This has important implications for the hypothesis of the FPS game acoustic ecology especially as it relates to player immersion within and participation in that ecology.

Curtiss, writing about sound practice and sound analysis in early Warner Bros. cartoons, suggests that such cartoon sound can never be indexical, it is either isomorphic or iconic; whereas live-action films tend to use both indexical and iconic sound, cartoons only use iconic sound "... indexicality is impossible in a cartoon" (p.202). An indexical sound is one having a direct, causal relationship between the sign and the signifier, the sound and the object. An iconic sound has no direct causal relationship to its object, it cannot have been caused by or sounded by the object seen on screen. Isomorphism (sometimes known as 'mickey-mousing') is less prevalent in live-action films and, given the dominance of dialogue in popular commercial film, indexical diegetic sound is more common than iconic diegetic sound. Comedy films have a long history of both iconic and isomorphic sound — one has only to think of the films of Jacques Tati for examples of iconic sound usage — and, as Curtiss himself points out, other film genres such as *film noir* have made use of isomorphic music (p.203). These are exceptions rather than the rule though.

For the Warner Bros. cartoons discussed in his article, Curtiss prefers to use the terms isomorphic and iconic rather than diegetic and nondiegetic claiming that the latter have little relevance where inanimate objects dance to a musical groove and where the musical beat and tempo are determined before ink is put to cell (a good example of a film's musical score bearing a strong relationship to, if not defining, the diegesis). He defines the two terms thus: "If isomorphic relations refer to those governed by rhythm and movement, then iconic relations pertain to analogous relationships between visual events and the timbre, volume, pitch and tone of the accompanying sound" (Curtiss, 1992, p.202). Here, isomorphism means same-shape whereas iconic is used in a Peircean sense as an analogy between the sound and the object.

In Curtiss' sense of indexical, it would be true to say that sound in FPS games is never indexical. There is no direct, causal relationship between the sound heard and the pixels on screen although, perhaps, a potentially specious argument could be constructed that sound is indexical based upon the gameplay where the sound heard has a causal relationship with the game engine that manages the gameplay. Discussing sound in *Half Life 2*, Breinbjerg suggests:

> In our natural way of listening sound is indexical. It points to the fact that a given event is taking place. Sound occurs only when materials interact [...] The interaction is to be understood as a source-cause relation [...] in which a sounding system (the source) resonates as a consequence of a given action (the cause), like when a hammer hits a bell (p.3).

He does not discuss this indexicality further in the context of game sound other than "we should take care in favoring the indexical nature of the sound event" (Breinbjerg, 2005, p.5).

Curtiss argues that cartoon sound can never be indexical since it is physically impossible for the animated objects on screen (at any stage of the production process) to have been the source of the sound (p.202). The same reasoning can be applied to sound in computer games. The actual 'event' is the execution of some game code. This does not make sound and is not an interaction of materials. The code is not the sounding system (the source), and the action (the cause), in this case, is the player navigating the code and triggering code execution. This is the only interaction here leading to sound and it is not an interaction of materials. It is clear, though, that the intention of the FPS game sound designer (where some semblance of a reality is being simulated) is not to provide iconic sounds but to provide sounds that have the *potential* to be (believed by the player to be) indexical if only the pixels on screen were rendered in a photorealist manner. In *Urban Terror* (Silicon Ice, 2005), no sound is synthesized and all sound samples are recordings of (or assembled verisimilitudinally to represent) the real-world objects depicted within the game (Klem, personal communication, 2004). When a shotgun is fired, the sound that is heard is a recording of a shotgun; when a wolf howls, the sound heard is a recording of a wolf howling. This indexicality of intention is always the case for sound in *Urban Terror*.

van Leeuwen states,

> [s]ound [...] is *designed.* It is no longer 'slaved' to what we see, but can play an independent role, just as many of the sounds in our everyday environment are no longer 'indexical', mechanically caused by whatever they are the sound of, but designed [...] [A]s I open a door, I may hear, not the clicking of the clock, but an electronic buzz [...] [M]uzak replaces the waiting tone of the telephone (pp.167—168).

Sound is also designed for the FPS game but, I suggest, in the FPS context, it remains indexical or takes on the mantle of a different indexicality. There is often a certain amount of processing applied to the recording before it is packaged with the game media, much of it to do with the processing of amplitude and frequency. Furthermore, some FPS games will process sound in real-time. There is, therefore, no such thing as the 'recording of a shotgun' within the FPS game context and, as stated above, the notion of an original sound (that is, the recording) is theoretically worthless in the context of FPS gameplay. Because the audio sample is caused to sound by the combination of the player's input and the game code *at that moment in time* and, furthermore, may have DSP applied depending on the context, the sound becomes indexical to the events being depicted on screen and thus indexical to the player's actions.

Indexicality of sound to the player's actions in the FPS game, married to an indexicality of intention, has importance to player immersion in the game's acoustic ecology through implicating the player in that ecology. The player's actions have a significant impact upon the acoustic ecology heard by that player (and other players in a multiplayer game) and thus the majority of sounds heard are a result of player action. Sound is indexical to the player's action through a process of cause and effect (I develop this further in chapter 7). This indexicality of sound to the player's actions may be usefully combined with the idea of the first-person auditor and strengthens the case for the hypothesis by being a close parallel to how humans both make sounds and hear those sounds in a real-world ecology.

5.7 Conclusion

Sound typically works together with image, with event or with both to create what may be viewed as one perceptual event. This is how sound works in the real world and how it may be assumed to work in the FPS game world.[31] There are some differences, notably that image in the real world is 3-dimensional yet image in the game world is merely an illusion of 3-dimensionality as represented on a 2-dimensional screen.

The theorizing of sound in film theory provides some useful terminology which, with various modifications suited to a different medium providing more kinaesthetic control than does film, may be used for the development of a conceptual framework to aid in supporting the hypothesis. Thus sound may be acousmatic or visualized sound, it may be kinaesthetically controllable by the player or not and it is possible to derive from it information about the visual environment of the game. Interactable sounds

[31] It is to be hoped that future empirical work will support this idea.

are typically associated with the event that object is responsible for while environment sounds are typically associated with images or locales (which may be off-screen and therefore implied). Importantly, much of the sound of the FPS game acoustic ecology is predicated upon player actions rather than images on screen and this is a significant factor in the immersion of the player in that ecology and, therefore, in the game world. Furthermore, it may be surmised that the relative proportion of kinaesthetically-controllable, interactable sounds to all other diegetic sounds is a clue to the level of immersion possible in the game. The higher this ratio, the more opportunity the player has to participate in the construction of the FPS game acoustic ecology and this, as explained in chapter 7, leads to greater immersive possibilities.

The model of the FPS game acoustic ecology may now be expanded to take account of the main points raised in this chapter — especially the effect of player input, causality/indexicality of sound, distinctions between acousmatic and visualized sound and the notion of synchresis.

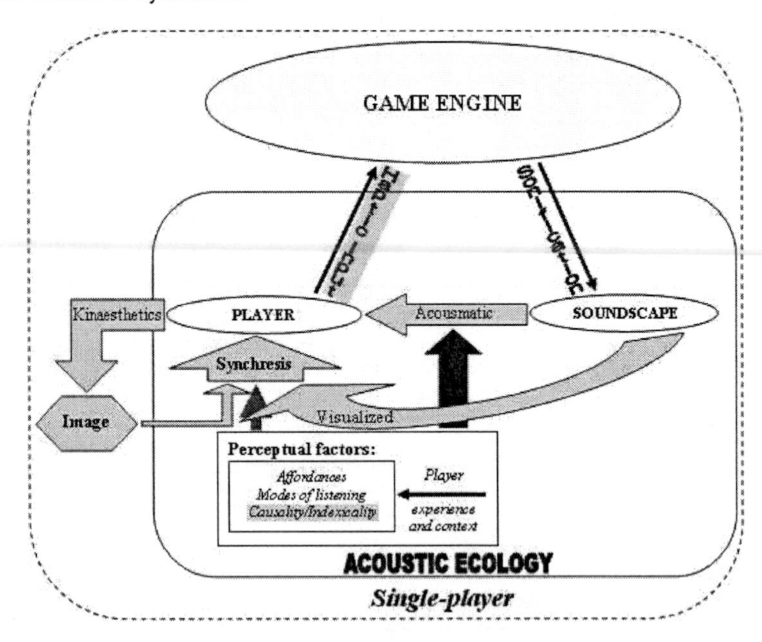

Figure 5.1. The FPS game acoustic ecology model accounting for the relationship of sound to image and game event (single player).

This now indicates that the triangular relationship between player, game engine and soundscape may be further explained by the process of haptic input. In other words, many of the audio samples in the game engine are triggered by the player's mouse and keyboard (or console) actions and these actioned samples (as explained in chapter 4) are sonified into sounds that appear in the soundscape. Causality/indexicality of sound is bracketed under perceptual factors because it affects the meaning players give to sound and, in achieving this, is itself combined with other factors (such as affordances and modes of listening) and all of these are affected by the functional parameters of sound as interpreted by player experience and context. Importantly, though, in the context of audiovisual FPS games, image, and the relationship between image and sound, is now accounted for through the notion of synchresis and the classification of sounds as acousmatic or visualized. Visualized sounds are synchretized with the image displayed on screen providing the one perceptual event as explained above. Acousmatic sounds do not go through this process of synchresis but serve to indicate the existence of objects and events in the game world that are outside the space and view represented by the 2-dimensional image. Furthermore, acousmatic sounds may become visualized sounds acting synchretically with image through the player's kinaesthetic action.

The theory of synchresis, as derived from film sound theory, provides an explanation for the perceptual centering of sound on image in the FPS game and this is developed in chapter 6 to account for the synchronization between various spaces in the acoustic ecology and the spaces of the game world as seen on the screen. However, synchresis does not account for the melding of a real 3-dimensional soundfield with an illusory 3-dimensional game world which the player may navigate because cinema does not have this level of navigational interaction. The notion of a first-person auditor, in which the player hears from within the (sonic) game world combined with the navigational listening mode, goes some way to filling this conceptual gap and this notion is further strengthened by the proposal that many sounds in the FPS game are indexical to player actions. The first-person auditor hears from within and is able to navigate the acoustic spaces of the game world, and these spaces are the subject of the next chapter.

Chapter 6

Acoustic Space

Introduction

Chapter 5, in part, dealt with the relationship between sound and image in the FPS game. The images of an FPS game help define spaces with which a player may engage and through which his character navigates. This chapter explores how the spaces in which the gameplay takes place are related to a variety of spaces represented by the FPS game sounds and outlines the consequences of this for the later discussion of player immersion and its role in a model of an acoustic ecology (chapter 7). Images in the game world are 2-dimensional and any 3-dimensionality they imply is purely illusory. Sound, though, can only exist in a 3-dimensional form and this, as argued in both this chapter and in the next chapter 7, is the reason why sound is of prime importance to the perception of spaces, both within the acoustic ecology and within the game world, and to player immersion in these spaces. An ecology comprises a set of spaces in which the components are contained or are active and this, I suggest, is the reason why spatiality is an essential requirement for any discussion of the FPS game acoustic ecology.

In answering the question *Does sound contribute to the spatial dimension of the game and, if so, how?*, the chapter provides further answers as to what the term acoustic ecology implies in spatial terms, how this relates to the game world and what implications this has for player immersion within and participation in that ecology. I suggest that approach to the FPS acoustic ecology may be understood as an acoustic space comprising a set of resonating spaces and paraspaces. 'Paraspace' follows Parkes and Thrift's (1980) terminology and so is a space which, within the acoustic ecology, describes and provides immersory and participatory affordances to do with location, time and other more cultural and social factors. These help to anchor the player in the perceptual reality (see chapter 7) of the FPS game world by providing the immersive cues that enable the player to become an integral part of the game's acoustic ecology, thereby affording the perception of that ecology and the components within it, and by inviting participation in the creation of that ecology.

Although Parkes and Thrift contrast paraspaces with universal spaces (which are dimensional or volumetric spaces that must be discussed alongside time), within the concept of the FPS game acoustic ecology, the word 'universal', where, in section 6.2, I suggest that there is more than one of these types of spaces, is fraught with difficulty and confusion.[1] Thus, I propose the term *resonating space*[2] because it implies a number of properties and functions which relate to the idea of an *acoustic* space. Although it may imply other quantities, 'resonating' has a particular affinity to acoustics and, in this sense, it also implies reverberation, localization and the notion of sound propagation from a source through air as the soundwaves expand in the

---

[1] Regardless of the notion of multiverses.

[2] I am indebted to Bevin Yeatman for the name.

volume of the containing or reflective spaces[3] represented in the images on screen. In other words, volume and time (no longer separate) are two important components of this resonating space and thus this spatial concept is admirably suited to discuss sound. Furthermore, in its recognition of volume, the resonating spaces of the FPS game acoustic ecology are a perceptual reification of the Cartesian coordinate system embedded in the game engine and, in its recognition of the temporal element, such a spatial concept recognizes not only the temporal dimension of sound but also the essential dynamic nature of the acoustic ecology — FPS spaces change.

Breinbjerg (2005), uses the terms *architectural space*, *relational space* and *space as place* (pp.3—5) when analyzing sound use in *Half-Life 2* (Valve Software, 2004). The latter two terms relate to the concept of paraspace and it is in this sense that they are discussed here. However, architectural space, while it does imply the notions of volume, containment and, in the context of sound, reverberation, is a problematic term in the context of modeling an acoustic ecology. There is no implication of a temporal property and, furthermore, it is explicitly tied to architecture, to human industry, and therefore makes no allowance for the fact that such a space may also be found in natural ecologies (sound can resonate through valleys and reverberate off mountains). Although a minor point, it is important to take into consideration the range of non-architectural constructions that may be found in many FPS games and this, the more neutral term resonating space does. Breinbjerg, however, does state that architectural space is a quantitative phenomenon while relational space and space as place are qualitative phenomena (pp.3—5) and this important distinction may be applied to the spatial terminology I utilize. Thus, resonating space is a quantitative space whose dimensions may be measured[4] whereas paraspace is a qualititative phenomenon recognizing, in this case, player knowledge and socio-cultural experience when defining the qualities of such a space.

This chapter, then, explores the import of sound to the perception of spaces and adds a further dimension to the conceptualization of the FPS game acoustic ecology. This exploration helps to provide further answers as to how and why players might become immersed within and participate in the acoustic ecology and so serves as preparation for chapter 7. The present chapter begins with an outline of the localization and reverberation of sound, as these are the main tools for the game sound designer with which resonating spaces may be designed, before proceeding to discuss resonating space, acoustic paraspace and acoustic time, and I conclude with a brief discussion of the FPS game soundscape serving to illuminate further paraspatial elements of the FPS game acoustic ecology.

---

[3] Although sound can certainly propagate through an 'open field' where there is no such containing space or reflective surfaces, such situations are extremely rare (so rare in nature that they may, in fact, not exist). Furthermore, due to technical constraints in the real-time production of graphical components during the gameplay, FPS games are often full of such spaces and surfaces.

[4] Not necessarily exactly, but it is possible to say, for example, that the volume represented by a resonating space is large, small or larger than that one. Propagation times may, however, be quite accurately measured from source to listener.

6.1 Localization and reverberation

The two qualities of sound which prove to be important in the perception of spaces defined by that sound are localization and reverberation.[5] The localizing properties of sound are evidence of the real 3-dimensionality of sound and reverberation is an effect of the dimensions and disposition of objects in the space in which it sounds as well as the material qualities of that space. This section, therefore, is a brief discussion of these two qualities as preparation for later, more focussed debate on the spatializing properties of sound as it might be understood in the FPS game acoustic ecology.

One of the key factors that must be considered in relation to a conceptualization of participation in the FPS game acoustic ecology is the ability to identify the direction a sound is coming from and, thereby, to determine the location of the sound source. This ability, if the sound source is unseen (perhaps behind the player's character), is the basis for kinaesthetic interaction with sound objects within the game (the possibility to acousmatize or de-acousmatize sounds as discussed in chapter 5) which itself is one of the gateways to understanding participation in the acoustic ecology. Additionally, localization of sound presupposes that there is a 3-dimensional space in which sound sources are positioned and thus the ability to localize sound in the FPS game is strong evidence of at least one resonating space extant in the game.

Localization of sound, disregarding synchresis and related audio-visual proximity effects, is mainly dependent upon the auditory system. Where the eyes provide stereoscopic vision for the perception of depth, the ears provide binaural hearing for the perception of location. The study of sound localization cues and localization perception constitutes a large part of the literature of both classical acoustics, ecological acoustics and the design of auditory interfaces and virtual reality.[6] The ability to localize sound is deemed to be of great importance for virtual reality or any system attempting to synthesize and implement an acoustic space or to represent some non-audio spatial data set in the audio domain. As discussed in chapter 5, where sound is juxtaposed to image, synchresis, or the audio-visual proximity effect, can have an influence on the perceived location of sound. However, in the context of an article on audio-only games, Röber and Masuch (2005) suggest that the "most important quality to explore acoustic spaces is the ability to determine the origin of sounds and speech" (p.3) and I suggest that this is particularly import in the 'hunter and the hunted' context of FPS games. While this work does not investigate audio-only games,[7] audiovisual FPS games generate sonic spaces (in addition to other forms of space) which the player's character inhabits and in which the ability to localize both looping sounds and discrete sounds is of great importance.

---

[5] There are other qualities too. Amplitude, for example, and the frequency components of sound are an indication of depth and the material qualities of the physical space in which the sound sounds. For the purposes of this brief discussion though, these are subsumed under the qualities of localization and reverberation. Additionally, strictly speaking, localization is not a quality of sound in the sense that it is a parameter of sound *per se*. Localization is something which affects the manner in which sound is sensed and perceived though, and it is for this reason that the more encompassing term *quality* is used.

[6] A larger treatment of the subject may be found in any good acoustics or psychoacoustics text book (such as Everest, 1984; Howard & Angus, 1996).

[7] Those interested in this may consult a range of articles and links at *Audiogames.net* (http://www.audiogames.net/)

There are several technical difficulties which limit the availability or effect of both localization and reverberant cues. The unknown factor of the player's use of stereo loudspeakers, surround sound systems or headphones presents problems for designing sound localization cues in FPS games. Nutating (of the head) will have some effect in both front and rear discrimination with surround sound speakers[8] but will only have a supererogatory effect for stereo speakers. Lateral localization will only occur on a small frontal plane when listening to stereo speakers and both speaker systems will experience similar difficulties in the provision of elevation localization cues while headphone usage negates the effect of nutation.[9] Given the almost infinite possibilities for the positional relationship between a player's in-game character and any one sound in the game, an infinite pre-reverberated number of variations of that one sound (such as a game character's footsteps in different locations, on different surfaces and at different speeds) cannot be added as part of the game data without presenting insurmountable problems of data storage (as noted in chapters 3 and 4).

However, despite these difficulties, the relevance of localization to FPS games should not be overlooked. This is particularly the case in those situations where the player utilizes the previously identified navigational listening mode (chapter 4) which depends for its effectiveness on the ability to locate the source of a sound. Certainly, an inability on the part of the game engine to encode the localization of an enemy's footsteps or an inability on the part of another player to perceive the direction and location of those footsteps will result in a very short in-game life for that player. Localization must be designed into the game and must be able to cope with a range of audio hardware situations. Whereas in the real world, a static sound source can be located by moving our heads or our bodies in relation to it, this is only an illusion in FPS games. The physical body, beyond some haptic hand movement (spasmodic or smooth depending on the run of the gameplay for that player) or whole-body reaction at an unexpected sound, does not move in relation to the sound sources. The illusion is given that the player's character moves within the game towards or away from static sound sources but, like the illusion that the sun arcs across the sky, it is only an illusion. Physical reality as found on Earth is reversed and static sound sources move in relation to the player who becomes a static figure back in its pre-Copernican position at the centre of the game universe.

Typically, environment sounds which are not kinaesthetically controllable by the player (chapter 5) and various globally sounded game status messages are not localizable (nor, given their function, need they be). For some sounds, localization may be treated as their most important function and for others, their low frequency, combined with the physical features and dimensions of the listener, means they are inherently difficult to localize. As Schafer (1994) puts it, in the context of music: "Localization of the sound source is more difficult with low-frequency sounds, and music stressing such sound is both darker in quality and more directionless in space. Instead of facing the sound source the listener seems immersed in it" (p.113). In this case, the very fact that such a sound is difficult to localize contributes to the perception of being immersed within a womb-like acoustic space.

---

8 Assuming the game engine fully supports the use of such a system.

9 The development of head-tracking devices, as described by Murphy and Pitt (2001), may well enable nutation to be coupled with headphone use if the tracking can also be mapped to sound localization cues as provided by the game engine.

While localization guarantees that there is a resonating space in which the sounds sound, it does not delineate the boundaries of that space. The relative intensities of different sounds or decreasing or increasing intensity of sound as the source object moves relative to the listener help to provide this. However, as a vital component in the perception of volume depth in the acoustic space, the reverberation of sound should not be ignored in this respect. Furthermore, reverberation helps to provide information about the sound source environment's material properties (cues to the game's paraspaces) and aids in immersion in the acoustic ecology by the process of externalization.

Shinn-Cunningham, Lin and Streeter (2005) suggest that "reverberant energy is necessary to allow accurate distance perception and to improve the subjective realism of a virtual auditory display". Whilst there is no discussion of what 'subjective realism' is, the desire to improve it may be a recognition that reverberation has significance in understanding of and contextualization within real-world auditory spaces. In any reverberant space, a significant cue for the distance a sound source is from the listener is the ratio between direct sound (sound which reaches the listener direct from the sound source without any reflection) and reverberant sound (sound which reaches the listener after being reflected off surfaces). The closer the sound source is to the listener, the greater this ratio is (that is, there is a greater intensity of direct sound compared to the intensity of reverberant sound) and the more distance there is between the sound source and the listener the smaller this ratio is (that is, there is more reverberant sound compared to direct sound). Therefore in media such as digital games, the addition of greater levels of reverberation, when compared to levels of direct sound, not only aids in externalizing the sound (when using headphones) but also perceptually distances the sound source away from the listener (an effect perceived over any loudspeaker or headphone system).

Shinn-Cunningham, Lin and Streeter further suggest that reverberation has a use in avoiding the internalizing effect of In-Head Localization (IHL). IHL is a subjective phenomenon where the sound heard over headphones is not externalized by the listener and appears to come from within the head. In part, this is due to the ineffectuality of nutating or otherwise changing the listener's physical position in relation to the sound sources. Wenzel (1992) suggests that IHL may be mitigated by the addition of environmental cues (pp.83—87). It is not clear what these 'environmental cues' may be but if, in addition to discrete sounds, we may surmise reverberation, then the results of Shinn-Cunningham, Lin and Streeter appear to support Wenzel's suggestion that sound is more likely to be externalized through the addition of such cues. Furthermore, placing the listener at the centre of a universe of externalized sound is a fundamental requirement for player immersion, a position strengthened by the identification of the first-person auditor in chapter 5. This immersion cannot happen if the acoustic ecology itself is immersed in the player's head through the process of IHL. Despite Schafer's hyperbole that the headphone listener *is* the sphere, *is* the universe of sound (p.119), the player should not *be* the universe but should be the *centre* of that universe and, therefore, *within* a sphere of moving sonic elements. With the player as first-person auditor, this universe of game elements does actually, in technical terms, revolve around the player — it is the game engine that rearranges these elements, both visual and sonic, around the static position of the character so creating the illusion of movement in a 3-dimensional game world.

One final role that reverberation plays in the creation of both resonating spaces and paraspaces within the FPS game is its betrayal of the surrounding environment's material and spatial properties. Long reverberation times[10] are dictated, in the main, by two factors. One is the volume of the physical space in which the sound sounds and the other is the absorbent qualities of the materials off which sound may be reflected (reverberation is the sum of all sound reflections as opposed to the direct sound reaching the listener straight from the sound source). Hence the long reverberation time of cathedrals which are very large, enclosed volumes and which are usually made of reflective stone or plasterwork.[11] Thus, in the FPS game, a highly reverberant sound is typically used to indicate that the player's character is within a large, reflective space as opposed to being in an open sound field (such as outside). The length of reverberation is also indicative of the virtual material in the game world off which the sound is apparently reflecting. In this case, it is not just the length but also the timbral quality of the reverberation — sounds reflecting off metal surfaces tend to be quite bright as opposed to sound reflecting off wooden surfaces for example.[12] In this way, reverberation helps create the perception of paraspaces in the game such as specific types of locales and buildings as indicated by their materials.

Modern FPS games localize almost all diegetic sound by placing each sound in a particular position in the audio field (stereo or multichannel) with respect to a central point in the game world which, for each player, is the player's character's location in that world. In other words, sound revolves around the first-person auditor (see chapter 5). The act of sonic localization creates the basics of a 3-dimensional acoustic ecology which (in addition to paraspaces which I discuss further in section 6.3) comprises at least two acoustic resonating spaces: the one that exists in reality and the virtual acoustic space that mirrors and expands in all directions the visual space seen on the screen. While the player is physically immersed in the former, the latter invites immersion in the game world by, among other things, allowing the player to kinaesthetically interact with the acoustic space. Reverberation, too has a role to play in the creation of immersive spaces in the game. It aids in externalization, depth and distance perception and, because any reverberant part of sound is founded upon the characteristics of that sound, it also plays a part in the causality and indexicality of that sound because, as discussed in chapters 4 and 5, a sound may be characterized by its source object and source action. Furthermore, as reverberation is dependent upon the physical characteristics of the surrounding environment, its materials and volume, it is, therefore, a significant factor in the perception of these materials and volumes. As Chagas (2005) states: "Reverberation becomes an instrument of the *deconstruction* and *re-construction* of space" and thus it helps to conceptualize the variety of resonating spaces and paraspaces that are found in a model of the FPS game acoustic ecology.

---

10 T_{60}, the time in seconds that it takes for the reverberation intensity level of a sound to decrease by 60dB.

11 Anechoic chambers are typically small and comprise highly sound-absorbent materials.

12 Metal usually absorbs low frequency energy to a greater extent than high frequency energy.

6.2 Resonating space

In the introduction to this chapter, I defined the term *resonating space* as an acoustic space comprising a volume and time in order to distinguish such a space from acoustic paraspaces which I discuss in section 6.3. This section, therefore, expands upon that definition and uses it to discuss those acoustic spaces differentiated in the FPS game ecology which are defined in terms of volume and time and which, therefore, provide spaces in that ecology for players to be immersed within. For this discussion I make use of the outline I have provided in section 6.1 in regards to the localization of sound and reverberation. I also make use of concepts I have discussed in previous chapters such as the navigational listening mode and kinaesthetics and synchresis (chapters 4 and 5). I argue that there are at least two resonating spaces to be differentiated in a conceptualization of the FPS game acoustic ecology and that the player is physically immersed in that resonating space which is real (and which exists in the user environment) and that this real resonating space acts in conjunction with the graphical spaces depicted on the screen to create the perception of a virtual resonating space which is a sonic reification of the spatial elements of the game engine. Finally, I provide new terminology with which to classify those sounds that contribute to the perception of these resonating spaces.

The FPS player's prosthetic limbs, extend away from the player from reality into a visual space that, in graphical terms, is a representation on a 2-dimensional screen of a 3-dimensional space. In visual terms, the space into which the player thrusts his prostheses is completely illusory; it makes use of Western artistic perspectivist conventions, scaling, simulations of parallax (for depth) and light and shadow to simulate a 3-dimensional space such as those found in reality. The game engine aids in this simulation, for example, by mimicking some of the effects of gravity and other aspects of the game such as 'solid' surfaces through which bullets cannot pass, or by the ability of characters to wander into doorways, around objects or up and down stairs.

But all of these aspects are merely simulations because their effects and actions cannot be sensed (bar some primitive haptic feedback device which smears with a broad brush the delicate tracery of tactility) but only perceived as something which is happening to a screen character we are asked to identify with. Artistic perspective and scaling is a simulation of objects receding in size towards the horizon coupled with decreasing discriminatory ability on the part of the graphics system. Parallax is simulated by the differential repositioning of visual objects on the screen as the character 'moves' in the environment such that objects nearer the player are shifted more than distant objects. This shift is usually lateral but in some cases, particularly where there is more vertical freedom of movement within the game, may also be vertical. Gravity in the game does not exert a force on the player who senses only the effects of reality's gravity. It is not the player feeling the effects of bullets ripping through flesh or suffering after a three story drop from a roof. These are vicarious pleasures derived from the simulations of the game.

Sound though is different. It is not a simulation of 3-dimensionality — it cannot be as sound must be heard in reality to be sound.[13] In other words, sound cannot be sound

---

[13] Sound is a waveform that propagates through air (typically) and, in so doing, it occupies an expanding volume. This waveform has the parameters of pressure and frequency and, as long as the pressure is above 20×10^{-6}Pa and the frequency is within the range c.20Hz to c.20kHz, the waveform can be sensed by the ideal human auditory system.

unless it exists and is propagated in a 3-dimensional space and that space is within the space of reality not the space of the game world — it can only be heard when the data encoding the sound is freed from the 2-dimensionality of the game by being sonified into reality. Whereas a representation of a street on the screen exists solely as a flat, 2-dimensional representation that is created from data stored within the computer (until the holodeck becomes an actuality), FPS game sound (indeed any game sound) must be audificated into an audible analogue of the audio data stored within the computer and waiting to be ejected into the 3-dimensionality of reality. Sound, therefore, possesses spatializing affordances which are quite distinct to those of image and these, I suggest, better fit sound for the purposes of conveying the spatial characteristics of the FPS game world to the player than the use of image. Because sound is 3-dimensional (indeed the only truly 3-dimensional component of the FPS game), I suggest that FPS game sound is the physical gate through which the player enters the virtual world of the game.

The creation of the resonating spaces found in the game makes use of localization cues, reverberation cues indicating room size, depth cues (such as direct sound to reverberant sound ratios) and other cues such as doppler shift indicating speed and direction of travel of the sound source (indicating, in other words, that there is a volume for sound source objects to move about in). As with the direct sound produced without spatial processing from the FPS game's audio samples, these spatial cues can only be heard as a propagation of a soundwave through 3-dimensional space. The conclusion therefore, is that whereas the visual aspects of an FPS game are a 2-dimensional simulation of a (imagined or real) 3-dimensional space, the sound elements are a (re)creation in 3-dimensional space of another (imagined or real) 3-dimensional space. To a certain extent, this (re)creation is only partially successful because of the separation of sound from the apparent cause (as seen on the screen). Sound that appears to be coming from the objects on screen actually comes from loudspeakers and this introduces some limits in our sensing and perception of the sound. These limits relate to the inability of the player's movement (the player, not the character) to have the same effect on depth and localization cues as would be expected were the sounds to issue from the environment outside the world of the game and the loudspeakers (see above).

The real resonating space issues out of the game world into the real-world space of the user environment through the agency of audification and this space provides the only form of physical immersion which digital games currently offer.[14] This real resonating space, then, is the physical doorway to the virtual world of the game. The second is the virtual resonating space which, as I suggest below, is a recreation of the spaces suggested by the image on screen that is mapped onto the image through a process of synchresis and, in so doing, both amplifies and confirms these suggested spaces. For the conceptualization of player participation in the game's acoustic ecology, the player must be understood to be immersed in that acoustic ecology and, here concentrating solely on spatial issues, I suggest that the player may be considered to be physically immersed in the former resonating space but is mentally immersed in the latter and that it is the conjunction between the former space and the images of the game world which brings this about. In becoming

---

Audio data stored by analog or digital means is not, therefore, sound; it is merely a sequence of digital 0s and 1s awaiting audification.

14 Disregarding a variety of arcade game frames which, in most cases, may be viewed as add-ons to the game.

immersed within the virtual resonating space (and therefore immersed within the FPS game acoustic ecology), the player is making use of sonic affordances (some of which pertain to such spaces) and uses listening modes (particularly causal listening but also navigational listening because the resonating spaces in the ecology are spaces which can be navigated).

Resonating space,[15] is defined by a number of environmental factors which are added to the direct sound that issues from the sound source and that therefore alter the sound that the listener senses and perceives. Such environmental factors include reverberation (direct sound is reflected off surfaces), distance of sound source from the listener (decreasing overall sound intensity and decreasing direct sound to reverberant sound ratio with increasing distance) and lateral, and to some extent vertical, position of the listener in relation to the sound source (localization). The final sound, therefore, is a result of direct and reverberant sound, from the stationary or moving sound source, modified and being changed in time by the effects of objects in the vicinity.

The perception of the sound that results from these environmental factors will be different for each person. There are several reasons for this. Firstly, every person perceives the same sound sensation differently because of different physiology and other causes such as the absorptive or reflective qualities of the clothing worn. Secondly, the sound sensation changes between person to person if, for example, person A is further from the sound source than person B or if person A is positioned in a more sound-absorbing area than person B. Thirdly, the perceptual and cognitive faculties of each person will categorize and provide meaning to sound differently. These differences may be manifested by varying intensities of sound at each person, different frequency spectra, different ratios of direct to reverberant sound, by different localization cues or by different responses to the sound — an ambulance siren will have different meanings and responses required for a car driver ahead of the ambulance than for a car driver behind it.

There is a significant difference between the acoustic spaces created in films and those created in FPS games that I have touched on before (chapter 5). In film, offscreen sound that "enlarges the film space beyond the borders of the screen" (Truppin, 1992, p.236) does not invite or allow exploration into that space because the film spectator is an immobile participant in the film's space unable to discover the offscreen sound source unless the director wishes it. With few exceptions, the source of a sound from an object not in sight in an FPS game can be discovered by turning towards the direction the sound is coming from (that is, altering the screen view) and, if necessary, moving (the character) towards it; a kinaesthetic interaction. Whereas we can only imagine "the reality [...] of a larger world stretching on indefinitely" beyond that which we see on the cinema screen (Truppin, 1992, p.236), in the FPS game we can explore the space, guided by our ears, and confirm the world implied by the 'offscreen' sound.

This mimics one of the functions of sound in reality that is not provided by film sound — in reality too, a person may actively investigate the physical space that is hinted at by 'offscreen' sounds — and in this way, in the FPS game, the provision of a teasing

---

15 Whether the real resonating space in which the player is physically immersed or the virtual resonating space in which the player's character is immersed.

acoustic space with New World promise[16] as an analogue to the acoustic spaces of reality provides some of the elements of immersion which I discuss in chapter 7. For example, gunfire at various points of the FPS acoustic space may be classed, at a first level, as perceptual sureties if the player is in the midst of a fire-fight. They may be classed as perceptual surprises of the attractor and connector variety if the gunfire, perhaps in conjunction with rapidly approaching footsteps, suddenly sounds to one side of a player in the lull of a battle — attractors because there is an action expected on the part of the player (perhaps turn and face the potential enemy or run away) and connectors because they help orientate the player in the visual space depicted on screen.

As mentioned in the introduction to this chapter, Breinbjerg posits three dimensions of space as defined by sound in the game (Breinbjerg, 2005, pp.3—5). Architectural space is defined by simulations of materials used and the construction and dimensions of the space and is a quantitative phenomenon. Additionally, he makes the point that such space must be sonically indicated by objects that are not part of the architecture but whose sounds are nevertheless affected by it (for example, the reverberation of footsteps in a large, cavernous hall as the hall itself makes little or no sound). Relational space is defined by distance and position of sound sources in relation to the listener and is subjective, non-quantifiable and dynamic — listeners hear differently, are in different locations and use different audio systems — because sound sources and the listener may move in relation to each other.[17]

The use of the term 'space' (architectural and relational) by Breinbjerg is perhaps misleading as it suggests at least two distinct forms of space relating to volumes which are perceived by the listener. In terms of FPS game sound design, the distinction is a valid one especially when viewed as quantitative and qualitative because the first is relatively simply implemented in code (a simple reverberation algorithm for sound reflecting off a brick texture in a large room is a relatively trivial matter to code) and the second is more difficult to design because of the larger quantity of unknown factors (for example, the position of the player in the game, the use of surround sound or stereo playback).[18] However, in terms of player experience, I suggest that it would be more correct to label these as spatial components. The sounds heard in Breinbjerg's relational space are affected by his architectural space which requires the provision of other sounds which act in conjunction with the architectural properties to create the resonating spaces of the game. Any such architectural space requires a listener to be present within that space in order to hear the sounds — if there is no sensing auditory system present, there is no sound. Placing a listener in the architectural space immediately creates relationships between that listener and the components of that space (in which the absorbing and reflecting listener is now a component) and so the two spaces that Breinbjerg proposes are simply two facets of the same form of space which I term resonating space.

Within the proposed acoustic ecology of the game, sounds that enable the perception

---

[16] This conceit is a comparison to the *terra incognita* of early cartographers and the sense of wonder and imaginative surmise that contributed to the age of exploration.

[17] As noted above, virtually speaking.

[18] Increasing computer processing power will mean that relational space in the game will become more quantifiable as game audio engines become more sophisticated allowing the fine tuning, for example, of sound to player location in the game.

of the game world's resonating spaces (whether they are pre-reverberated audio samples or the sound heard after processing by the game audio engine), will henceforth be called *choraplasts* to indicate their ability to fashion the perception of a resonating space.[19] Sounds become choraplasts through the use of techniques such as localization and reverberation thereby providing the aurally perceived parameters of height, width and depth for the resonating space.[20] Components of sound which indicate temporal paraspaces will be analyzed separately in section 6.4, but it must be stressed here that any resonating space created by sound not only has volume but also has time. For the conception of FPS game sound as an acoustic ecology, this dimension is important as it is through time that sound sounds and over time that the ecology changes (and time also has importance in the context of telediegesis, an instance of player relationships in the game, which is discussed in chapter 7). This change reflects the integration and participation of players in the conceptualization of the acoustic ecology because it is an expression of the relationships, based on sound, between players and between players and the game engine.

It was suggested previously that there are at least two resonating spaces in the FPS game. The first, in which the player is physically immersed,[21] provides an example of sensory immersion in which the sounds in the user environment are overidden by the sounds of the game environment (Ermi & Mäyrä, 2005). The second, which sonically recreates the 2-dimensional screen image in three dimensions, is an example of perceptual immersion and, comprising as it does a number of perceptual immersion 'hooks', serves to lure the player into the game world inhabited by his character. Although I discuss these immersive factors in greater detail in chapter 7, here I would like to suggest that the real resonating space is mapped onto the game world by a process of synchresis and that this process is what creates the perception of the virtual resonating space. This is a type of meta-synchresis in which the entire soundscape (consisting of many individual sounds and which exists in a different location to the screen within the player's user environment[22]) is perceptually mapped not only to the image on screen but to the entire surrounding locale of that particular point in the game world (including the parts which are unseen). This meta-synchresis provides the affordance by which synchresis may then occur between individual sounds within the acoustic ecology and individual objects which are seen on the screen.

This section, then, has explored the idea that sound has a role in the creation of at least two resonating spaces in the FPS game, that these spaces physically immerse the player in one acoustic space and invite perceptual immersion in the other and that meta-synchresis, whereby the real resonating space is perceptually mapped to the game world despite being in a different physical location, is the process by which the virtual resonating space is perceived. Additionally, as part of the development of the conceptual framework with which to analyze FPS game sound, the proposed the term choraplast is useful for accounting for those sounds having the function of creating resonating space.

---

19 This has resonance with Kristeva's (1984) use of the term *chora* as a mobile and ephemeral articulation giving rise to, but preceding, spatial geometry (pp.25—26).

20 Note that the spatializing functions of sound including choraplast and others I define below (topoplast, aionoplast and chronoplast) are not mutually exclusive — any sound may have one or more of these or other (see chapter 4) functions.

21 In the case of headphone use, one might say physically encased.

22 A further set of resonating spaces and paraspaces.

Schafer notes that sound can be used to define a space that does not necessarily have any visible boundaries. In this case, sound is typically used to mark territory; for example, the territorial calls of birds or the sound of a parish church bell marking the purview of the parish. For Schafer, this territorially spatial use of sound is fundamental to many species and "[t]he definition of space by acoustic means is much more ancient than the establishment of property lines and fences" (p.33). On the subject of church bells, he states that it is a centripetal sound that draws communities together (Schafer, 1994, pp.54—56) and, later, gives the siren as an example of a centrifugal sound expelling people from the vicinity (Schafer, 1994, p.178). This is sound defining a social place or locale, a paraspace, and is the subject of the next section.

6.3 Acoustic paraspace

In addition to defining resonating spaces, and potentially in conjunction with vision, sound defines acoustic paraspaces. The conceptualization of the acoustic ecology of the FPS game may be viewed as a particular form of paraspace in that it is a space which enables social interaction between players, on the basis of sounds, and between players and the game engine. Here, though, the emphasis will be on the perception of locational paraspaces; that is, those paraspaces that indicate particular locales within the game world and are characterized, in the main, by parameters other than volume. In the process of doing this, new terminology will be added to the conceptual framework in order to classify those sounds possessing the function of perceptually recreating the FPS game's locational paraspaces.

Location may be general or it may be specific. Thus, to use my previous example from the real world, the sounds of traffic (car and bus engines, honking horns, for example), the footsteps of pedestrians on a hard surface, and muzak issuing from opening doors indicate that the location is a row of shops lining a street in a town or city. If the listener also hears the bells of Big Ben and a particular language (English) and accent (Cockney), this marks the location specifically as London. This may be compared to the locations found in FPS games although typically with less specificity.[23] For example, the Bach organ fugue and the birdsong audio samples in the *Abbey* level of *Urban Terror* (Silicon Ice, 2005) suggest that the location of the level is that of a Christian religious place in a peaceful countryside setting.

These characteristic identifying sounds may derive from the sound as it propagates direct from the sound source, they may be a combination of the direct sound and other factors such as reverberation or echoes or they may be solely from this reverberation. In the first case, an example would be the sound of seagulls and waves lapping on a beach marking the place as a marine littoral. An example of the second case would be the echoing sound and significant reverberation of a raquet hitting a ball and the ball hitting a wall indicating a squash court. The third case is exemplified by the long reverberation following any sound sounding in churches of the Gothic and Romanesque variety (as used in the Bach organ fugue in the *Abbey* level of *Urban Terror*). Sometimes, it is the absence of reverberation cues that hints at the place although sound unaffected by the properties of the place in which it sounds is rare in nature — speech in a flat, featureless landscape is probably as

---

23 Suggesting that locational specificity may be increased through a combination of sound and image.

close as one can get to it[24] — but the artificial example of such a place would be an anechoic chamber. It is also quite rare in nature to hear reverberant sound without its preceding and accompanying direct sound (this is easily produced artificially and is a favourite trick of sound engineers working in popular music) though an example may be a volcano exploding and echoing around the world.

Writing about film sound, Chion identifies sound types that help define a setting of a film (in this case he refers to location as 'space' or 'locale'). These he divides into discrete sounds emitting infrequently and more pervasive, permanent background sounds; thus "elements of auditory setting [which] help create and define the film's space" (Chion, 1994, pp.54—55). They may be synchronized to specific sound sources on the screen or not and his examples include a dog barking or an office telephone ringing. What Chion calls *ambient sound* is "sound that envelops a scene and inhabits its space [...] birds singing, churchbells ringing. [Although lacking a visual source they] identify a particular locale through their pervasive and continuous presence" (p.75). Unfortunately, there is some confusion in his terminology because a sound that is an element of auditory setting "inhabits and defines a space, unlike a 'permanent' sound such as the continuous chirping of birds or the noise of ocean surf that *is* the space itself" (Chion, 1994, p.55). For Chion, both types of sound *inhabit* the film's space and both define and identify that space or place, yet he first states that birds singing do not do this because they *are* the space and then later gives the same example of birds as an ambient sound *defining* and *inhabiting* the space. Regardless, there is a recognition that sound in film can be used to identify and define a location whether that location is a general or a specific location.

The same recognition may be applied to some of the sounds in the FPS game. The location may or may not have an equivalent in reality; the sound of Big Ben chiming in a level that represents London is an example, the sound of traffic in an urban environment is a more generic example and sundry more imaginative sounds indicating fantastic locations[25] are examples of locations having no correlate in reality.[26] Such sounds are interpreted via cultural experiences or (and sometimes in addition to) the experience and training gained in playing the particular game or level as in the last examples above. There is no appropriate term classifying such sounds and so here they will be termed *topoplasts* because of their ability to signify a place or location.

FPS game sounds that help create locational paraspaces are usually environment sounds or pervasive keynote sounds (for example, continually ringing church bells indicating a church in the vicinity or singing birds and the continuous sound of wind rustling through leaves indicating a forest). But they may also be more discrete sounds or signal sounds (which I discuss in section 6.5) that players actively attend to but, nonetheless, because of the identifying properties of the sound, help to

---

[24] Other than screaming while falling out of an aeroplane.

[25] During the design of the *Quake III Arena* (id Software, 1999) *Grim Shores 3: Atlantis* level (Grimshaw, 2001) (which was set in an underwater Atlantis complete with hieroglyphs and statues of Egyptian deities), my chosen ambient sound for a certain dark subterranean chamber was a susurration of voices created by recording myself chanting in Italian (an appropriately sibilant language), reversing the audio sample, multitracking it back on itself with a range of delays, applying a long, dark reverberation and finally discarding the direct sound in favour of the reverberant sound only. By no means was the speech intended to be intelligible despite being recognisable as some form of vocal utterance; it was designed for affect and to acoustically reify the volume of the chamber depicted on screen.

[26] Although they may well have cultural resonance, be re-imaginings of historical or mythical places or be interpretations of locations imagined by popular convention.

perceptually define that particular location. These types of sounds may also have the function of aiding in the creation of resonating spaces if the game engine processes them with cues indicating localization, reverberation and depth. Such sounds are able to function in this manner within the game because of the raw (that is, not processed in real-time) characteristics of the audio sample (for example the metallic ringing that indicates the tolling of a bell) or because the raw audio sample may be processed to serve as an indicator of place (by adding, for example, a long reverberation to indicate a cavernous, reflective place (such as a cathedral) which also serves as an indicator of resonating space.

It is important to note then, that a sound is not limited to just one of these spatial functions. Any sound may have a combination of functions which, furthermore, may include Ballas' functions (as defined in chapter 4). Additionally, understanding the paraspatial locational properties of a sound is often a matter of socio-cultural experience (as discussed in chapter 4). The sonic example of the ringing church bells given above may have little meaning to those who have never heard such a sound before but will have a host of connotations (perhaps different for different listeners) beyond the specific denotation to those who are familiar with the sound.

The game world of the FPS game is a compilation of acoustic locational paraspaces beginning with the general setting of the game level (docks, for example) and comprising various locations within that (quayside, warehouses, railway yard, for instance). Such paraspaces are contained within the real resonating space and, like the virtual resonating space, are meta-synchretically mapped onto the images on screen to become part of this virtual space. Sounds with a topoplastic function are key to enabling the perception of such locations and may work together with the image on screen for this purpose or, as 'off-screen' sounds, may function alone in guiding the player to a perception of locations beyond the limits of the screen.

6.4 Acoustic time

It was pointed out at the start of this chapter that time is a fundamental component of any discussion of resonating space or acoustic paraspace. However, simply noting this does not describe how the temporal components of sound in the FPS game acoustic ecology work with other sonic parameters towards the creation of these types of spaces. These components should be analyzed separately (as per choraplastic and topoplastic components) before being brought together in chapter 9 for a discourse on the acoustic ecology of the FPS game. This section, therefore, discusses parameters of sound which contribute to perceptions of time in the game and defines further taxonomies for the conceptual framework in order to describe the temporal functions of FPS game sounds. Some of this discussion compares perceptions of time in films to perceptions of time in digital games following the example of writers such as Wolf (2001).

Chion suggests that sound in film can have a temporal function when a montage or sequence of, at first sight, unrelated images is presented on screen. Without sound, these images can be read as either synchronous or sequential. However, "the addition of realistic, diegetic sound imposes on the sequence a sense of real time [...] a sense of time that is linear and sequential" (Chion, 1994, pp.17—18). Additionally, because sound is 'vectorized', that is, it has a definite start, middle and end, sound with moving images imposes a "real and irreversible time" — although many sequences of images are reversible without being noticeably reversed, sound in a reversed film always discloses that the film has been reversed (Chion, 1994, p.19). This is an important point, sound with moving images indicates that real time[27] is passing by and progressing forwards and this also relates to the progression of the diegetic time of the medium.[28]

As far as concerns the audio and visual reproduction systems of films and digital games, there are inherent technological differences between the two media affecting the perception of time. Staring at a 'still' image on the cinema screen (where neither the camera nor the subject move yet the film continues), the impression of time passing is cued by grain, flicker and other anomalies in the image such as scratches or hairs moving between the projector's light and the film screen. There is no equivalent in digital games (or, for that matter, in digitally produced and reproduced films). As Wolf states: "[G]rain, hiss and flicker are nondiegetic indicators of time passing [...] video games usually do not have the same nondiegetic indicators of passing time" and other elements, such as sound, must be added to indicate this (pp.79—80).

Some sounds have a function of marking and segmenting time in the present (whether 'present' in reality or the 'present' of film or game time). For these, there are a range of sounds or sound types such as clock-tower bells, marking the hours or fifteen minute divisions of the hour, or the ticking of a watch marking seconds.[29] These types of sound tend to be rarely used in FPS games other than as a musical or chordal flourish or similar when the gameplay ends or starts, and they segment time into large chunks relating to the completion of a level and the start of the next. An example of a shorter division of the gameplay time is found in *Quake III Arena* where various powerups are set to respawn at set time intervals at which point they will be signalled by a unique sound; the astute player will note the passage of time by such sounds positioning himself to take advantage of the powerup right at the respawn time.

For a sound which affects the perception of chronological time, its most important property is its macro-temporal property *viz.* its amplitude envelope. Thus a sound may have a *slow* attack, a *long* sustain and a *quick* release; all perceived and measured in terms and units of time and therefore having an effect upon the perception of chronological time. Additionally, the rate of repeating of the sound will potentially have a similar perceptual effect. For a sound having functions to do with the perception of resonating space and the paraspaces described above or historical timeframe, its micro-temporal properties come to the fore as these will be perceived less in terms of time units but rather in terms of their effect upon the density of reflections that make up reverberation, for example, or the timbre and spectral fluctuations that result from changing temporal relationships between the partials of the sound.

Sound also has the ability to indicate a point or period of time in the past, present or

---

27 The perception of 'real time' is used advisedly and is not necessarily a quantifiable amount as anyone who has lost themselves in a digital game only to later emerge discovering that several hours have flown by in the real world would recognize.

28 With the exception of flashbacks in film.

29 Heard less and less today.

future as pointed out by Schafer quoting line 501 from Book II of Virgil's *Georgics*: "Such was the life that golden Saturn lived upon earth: Mankind had not yet heard the bugle bellow for war, Nor yet heard the clank of the sword on the hard anvil". Hence certain sounds are redolent of particular historical periods; mechanical sounds, particularly metallic sounds, are indicative of the industrial and post-industrial ages and, by the same token, the absence of metallic sounds such as the 'clank of the sword on the hard anvil' can be indicative of the pre-Bronze Age period. The addition of some sounds, such as the sound of a jet aeroplane in a game environment meant to represent the mediæval period, can be anachronistic. In addition to the use of particular sounds, the density, loudness and frequency of occurrence of sounds can also indicate the time period — it is a common modern plaint that the background sound levels to which mankind is now exposed have increased significantly.

A final temporal function of sound, in conjunction with visuals, is one that indicates distance (and in this, it also has an effect on the perceived dimensions of the resonating space). Many of us are familiar with using the time between seeing a bolt of lightening and hearing the clap of the thunder to judge how far away the storm is. Because light at these small distances is perceptually instantaneous with the event, the distance of the storm is calculated from the speed of sound — as sound travels at c.344ms^{-1}, a delay between sight and sound of three seconds indicates a distance of approximately one kilometre. However, there are no FPS games which significantly use this phenomenon to indicate distance; this is strange as it would be a simple matter to code.[30] This flattening of time, or discarding of one form of distance perception cue, is an unusual feature of FPS games. It is difficult to justify why this should be the case especially if the rules of immersion require the provision of expected cues (see chapter 7) or if the FPS game designers are attempting simulations of aspects of reality. It may be an oversight brought about by a subconscious acceptance when designing the game that, despite the artifice of perspective and parallax in emulating visual depth, objects on the screen really are all on the same plane just a few centimetres in front of the player's eyes. It may also be that this oversight is due to a limitation of the current technology and number of audio channels available or it may be a desire to reduce real-time computer processing requirements. Perhaps there is a recognition that digital games should not be overly complex and therefore difficult to engage with and learn at first exposure or perhaps there is a requirement for an immediacy of the effect of in-game actions complicated by multiplayer configurations bringing the added complication of network latency.[31]

Because sound has time as a function of itself, that is, sound is heard in and may therefore be analyzed in terms of time, all FPS sound indicates the passage of time. Analyzing sound in this manner is an indication of the passage of real time where a sound has a certain length from start to finish or has a certain frequency (that is, cycles per second). There is also, however, a game time that is not always synchronized with real time if the game designer so wishes it and, additionally, there

---

[30] The relative size of an object displayed on screen is already derived in the code by its distance from the player's character so the same parameters used to calculate this can also be used to calculate and introduce an appropriate delay.

[31] Chapter 10 touches upon the necessity (or not) of emulating real-world ecologies in FPS games rather than merely simulating them.

are other perceptions of time, some more individual than others.[32] The ability to hear sound requires a length of time and so sound is indelibly linked to the passage of time. Furthermore, sound can also reference a certain point in time and, in this case, cultural and societal experience is required (chapter 4). For example, the modern industrial or post-industrial age may be marked by a plethora of cyclical mechanical sounds whereas the pre-industrial age is marked by an absence of or a decrease in the frequency and density of such sounds. Sound serving as indicator of era is, in many respects, similar to the use of sound as an indicator of location in that the characteristics of a sound or a group of sounds within the FPS game reference a particular period. Thus, a mediæval period (particularly for combat games such as the FPS genre) may be indicated by sounds such as galloping horses, the ringing clash of swords or the reverberance of a Romanesque cathedral as it affects plainchant for instance. Sounds that influence the perception of the passage of time are here termed *chronoplasts* and sounds which create the perception of the temporal period an FPS game is set in are termed *aionoplasts.*

Sound also serves to fit the playing of an FPS game level into a temporal container with beginning, middle and end (not to be confused with the classic three-act restorative narrative structure).[33] This may be defined by the density of sound (increasing as players hunt out and fight each other and decreasing as characters die and await respawning), it may be defined by particular types of sounds (sounds indicating the beginning, middle and end of the level) or by other methods depending upon the game engine and mode of play. This is clearly indicated by the graphical sound wave presented in *Figure 6.1* which describes the course of the gameplay of a free-for-all *Urban Terror* match with eight characters:

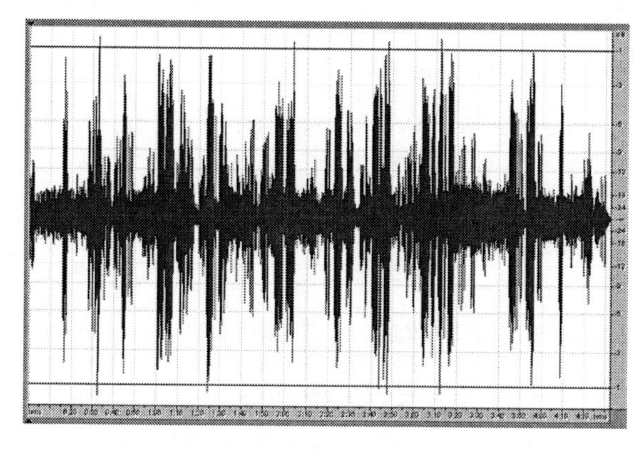

Figure 6.1. A visual representation of an *Urban Terror* soundscape over the course of a level.

This represents the sound heard by one player as he moves through the game hunting out and killing other characters or dying himself. The higher amplitude sounds represent the firing of guns at close quarters (the loudest are the player's own weaponry) and the points of increased density of such sounds are indicative of bouts

---

32 My personal experience is that real time tends to pass more quickly when playing a particularly involving FPS match; whenever I wish to kill time, I play an FPS game.

33 There may, of course, be other indicators of the game's timing such as game status messages.

of fighting interspersed with quieter periods during which the player is resting and bandaging wounds, waiting to respawn after in-game death or is moving towards the sound of distant activity in order to engage with the enemy once more.[34]

Sound, therefore, has an important function in FPS games in setting the period or age the action takes place in and in marking the passage of time — of indicating that objects are in motion and that the gameplay is proceeding. In chapter 1 it was stated that ecology was a preferred term to environment when describing the totality of FPS game sound because, among other reasons, ecology encompasses the notion of time, of player activity within the game and that the acoustic ecology will change over the duration of the gameplay. The role of sound in the perception of FPS game time, whether that game time is a progression of time or is a particular period of time, should not be overlooked as it is intimately integrated with the acoustic ecology of the game.

Having discussed and defined a range of acoustic spaces to be found within the world of the FPS game, I now move on to a discussion of the FPS game soundscape. Thus far, the discussion of FPS game sounds has revolved mainly around individual sounds. Here, the discussion revolves around sound in the context of other sounds and to do this I make use of Schafer's concept of the *soundscape.* This serves two purposes. It provides a further viewpoint of the acoustic spaces available in the FPS game acoustic ecology that is not opposed to the definitions I have provided above but, indeed, serves to complement it (particularly in the case of paraspaces). Additionally, it serves to prepare for a fuller account of the soundscape as a major component of the FPS game acoustic ecology in chapter 9.

6.5 The FPS game soundscape

The writings of Schafer provide another method of analysis for choraplasts, topoplasts, chronoplasts and aionoplasts which, because the writings deal with real-world acoustic ecologies, prove to be appropriate for use in the context of the FPS game acoustic ecology. Thus Schafer has theorized the soundscape in nature; that is, naturally-occurring soundscapes as opposed to the artificially created soundscapes found in cinema[35] and this is the subject of this section.

Some of the features of Schafer's soundscape are: *keynote sounds* which are ubiquitous and pervasive background sounds that are not always consciously attended to; *signal sounds* which are sounds in the foreground and that are consciously attended to and *soundmarks* (analogous to *landmarks*) which are sounds peculiar to that particular soundscape and help identify the place. It would seem obvious that these relate to Gestalt theory in the psychology of visual perception where a keynote sound forms part of the *ground* and signal sound is the *figure.* However, Schafer does not draw the parallel in part, perhaps because for him, a signal sound is not given mass or outline by the keynote sounds. Keynote sounds outline and give mass not to other sounds but to those living amongst them; humans

---

[34] It is interesting to note that, in this case, the period cannot be inferred from the visual representation of the soundscape in the same way that the graph above demonstrates the passage of time.

[35] Although film-makers such as Renoir stressed the importance of the use of naturalistic cinematic sound, nevertheless, such sound is still mediated and processed by recording and reproduction technology and so acquires an artificiality.

are characterized by their acoustic environment (Schafer, 1994, pp.9—10).[36] However, here I would like to initially suggest that, in similar manner, FPS players are shaped (or at least their game actions are shaped) by that part of the FPS game acoustic ecology which is formed by sounds triggered by other players or the game engine — in cases where the player himself triggers sound, it is he who has a formative role in the acoustic ecology. It is still the case, as suggested in chapter 5, that most sounds in the acoustic ecology are predicated upon player actions. (These relationships are returned to in more detail in chapters 7 and 8.)

Schafer's use of the term soundscape is analogous to the use of the term landscape but uses sound instead of image to describe the environment. For Schafer, that environment is a natural one but there is no reason why the concepts and terminology he defines cannot be applied to the virtual environments of the FPS game. A taxonomy of the FPS game soundscape, as part of the conceptual framework for an analysis of FPS game sound, would therefore be one that is divided into keynote sounds, signal sounds and soundmarks.

There is some similarity between the term keynote sound and the term ambient FX/sound as used by many film sound theorists (Chion, 1994 for example) or by those writing about digital game sound such as Folmann (n.d.). A keynote FPS game sound serves to provides sonic ambiance and to help in the definition of the space and location depicted on the screen but, usually, provides no further information to the player about interactable objects or actions. In other words, such a topoplastic sound helps set the scene of the FPS game level and has some similarity to the concept of a perceptual cue (see chapter 7). There are many examples of such sounds that may be inserted by the game designers as part of a game locale (birds singing in a forest, traffic sounds in an urban environment for example) or that may be associated with particular in-game objects (the sound of the fan in *Quake III Arena* for instance).

Where a keynote sound is part of a locale, it very often derives from a source object or action which cannot be seen or otherwise perceived and which, therefore, is merely implied (in other words, it is an acousmatic sound). This is often the case with pervasive traffic sounds in FPS game levels set in an urban landscape where no moving traffic may be seen. In reality, traffic is a common keynote sound as is the case while I am sitting here writing this work — the distant sound of unseen traffic reminds me that I live in an urban setting.[37] Where a keynote sound is associated with a specific object within the game environment, such objects are usually non-interactable, as is the case of the fan mentioned above. Where a sound derives from an interactable object, then it pays to attend to that sound in which case it can no longer be classified as a keynote sound — the *Quake III Arena* fire sound is a good example because it warns the attentive player of the dangers of wandering too close to the flames.

A signal sound is a foreground sound and therefore, unlike the keynote sound, is

---

[36] Describing the harmonics formed by the hum of electrical equipment in a street (such as street lighting and electrical signs) in 1975 in the Swedish village of Skruv, Schafer points out that they were found to form a G# major triad which became a G#7 chord (a dominant 7th) with the addition of a F# from the whistles of passing trains (p.99). If humans are defined by their acoustic environment, it is a shame that Schafer does not note a higher than normal incidence of stress and nervous disorder among the villagers from life lived in a state of continual cadential tension.

[37] Although, like Schrödinger's cat, the mere act of noting such keynote sounds, thereby consciously attending to them, will, it can be argued, negate its classification as a keynote sound.

consciously attended to. It is usually the case that these form the majority of diegetic sounds in the FPS game and they include sounds such as gunshots, footsteps, team announcements or any other sound which may be viewed as a perceptual surprise (see chapter 7) helping to immerse the player within the game.

Depending on the player's state of perceptual readiness (chapter 4) or the virtual distance from the sound, any FPS game sound may be classed as either a keynote or a signal sound. At the first playing of a game, the inexperienced player may consciously attend to all sounds and, therefore, all sounds may be viewed as signal sounds in this instance. The experience of playing the game later leads the player to classify sounds as those worth paying attention to and those not worth paying attention to — in other words, the player derives his own classification of signal and keynote sounds which develops with the experience and knowledge gained while playing the game or elsewhere (chapter 4). The hierarchy of affordances offered by sounds is adjusted by the player. Additionally, sounds which may be readily classed as signal sounds may also be heard as keynote sounds as a function of the character's in-game distance from the sounds. Thus an explosion or gunfire close to the player may be a signal sound but beyond a certain distance (and consequent attenuation in level and, if the game audio engine allows it, changes in the direct sound to reverberant sound ratio) may be viewed as a keynote sound because its source object or action is judged to be too far to be an immediate threat and the sound, therefore, becomes part of the game's ambient sonic experience.

A soundmark FPS game sound is an identifying aural feature of the game environment. Such sounds may be either keynote or signal sounds but are always dependent upon the player's prior experience of that sound. It may be that one sound or an entire set of sounds that are particularly identifiable or possibly unique to a certain FPS game becomes a soundmark for the entire game. The experienced FPS player will be able to identify FPS games within his ken by listening to a recording of the FPS gameplay soundscape or, in some cases, one specific sound. Furthermore, within the game, certain sounds can function as soundmarks dividing the game level into zones. In some cases (this is analogous to the potential use of landmarks), this can function in combination with navigational listening (chapter 4) as an audio beacon guiding the player around the gameworld from locale to locale or from action to action (for example, the organ music in the *Abbey* level of *Urban Terror* or the sound of gunfire in the distance).

Using Schafer's soundscape terminology, then, it is possible to classify all FPS diegetic sounds as either keynote sounds or signal sounds (a dynamic classification) and any such sound may be a soundmark or not. While such terminology is neutral with respect to the choraplastic, topoplastic, chronoplastic or aionoplastic properties of sound, because a soundscape is the aural equivalent of a landscape, the terminology carries with it connotations of resonating spaces (with their relationship to Gestalt theory) and, particularly, paraspaces which may, for example, be defined by territorial sounds, centrifugal sounds or centripetal sounds. These concepts prove valuable in formulating the conceptual framework of the FPS game acoustic ecology.

6.6 Conclusion

Whereas image in the game makes use of artistic conventions, artificial parallax and scaling in order to represent a 3-dimensional world on a 2-dimensional screen, sound alone is in three dimensions and, for this reason, is the primary perceptual enabler of the illusory space represented by the image on the screen. This real resonating space is audificated by the game engine from the audio samples stored on the installation medium. A second virtual resonating space is perceived through the meta-synchresis that occurs when the real resonating space is mapped to the image provided on screen. This is a particularly important affordance offered by the real resonating space and, in this way, the real-time-created resonating space provides the affordance of believing that the objects on screen are occupying or moving within the virtual resonating space (perceptually created by the FPS game) rather than within the plane area defined by the four sides of the monitor. These two spaces continually morph around the player adapting not only to the player's actions but also to other players' actions who are within virtual hearing range (a range which is dictated by the game engine). By doing so, they continually provide new affordances for player perception and response and this, combined with the afore-mentioned meta-synchresis, is a key perceptual hook used to immerse the player within the FPS game world.

In addition to resonating spaces, FPS game sounds also create a set of paraspaces of which the more relevant to the hypothesis are locational paraspaces and temporal paraspaces (there are two important ones — those contributing to perceptions of temporal period and those contributing to temporal progression). Furthermore, the functional ability of sound to provide spatial information may be combined such that any one sound may have, for example, both a resonating spatial function and a locational paraspatial function. The last section of the chapter discussed the soundscape of the FPS game not only as preparation for a fuller discussion of it in chapter 8 but also to propose complementary framings of the spatial functions of sound, particularly those which relate to paraspaces.

Pursuant to the aim of devising a conceptual framework for the establishment of the FPS game acoustic ecology, four terms have been created to describe sounds with a spatial function. These are choraplast (a sound whose function is to contribute to the perception of resonating spaces), topoplast (a sound having a locational, paraspatial function), chronoplast (a sound affording the perception of time passing) and aionoplast (a sound serving to set the game world in a particular past, present, future or more immediate timeframe). These terms are not mutually exclusive; a topoplastic sound may also have an aionoplastic function too, for example. Additionally, following soundscape terminology, sounds may be described as keynote or signal sounds and either one of these terms may be further classed as a soundmark. A classification of sound as keynote or signal sound or as soundmark is independent of its function as choraplast, topoplast, aionoplast or chronoplast.

Following the points raised in this chapter, then, the model of the FPS game acoustic ecology may be further modified as demonstrated in *Figure 6.2*.

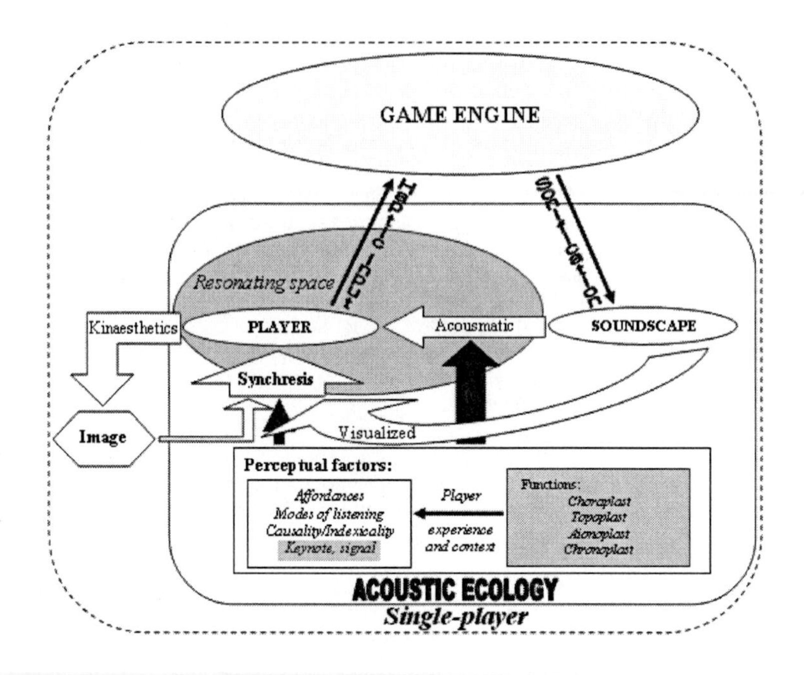

Figure 6.2. Spatial factors and soundscape terminology added to the model of the FPS game acoustic ecology (single player).

Figure 6.2 shows the player enveloped in the real resonating space that is created by the soundscape. Terminology derived from soundscape theories are present as perceptual factors. Furthermore, affecting the meaning of visualized and acousmatic sounds, the spatial functions of sound are mediated by player experience in their effect upon the player's hierarchical categorization of or use of perceptual factors such as affordances, modes of listening, keynote or signal sound and the causality/indexicality of sound. There are still a number of elements missing, though, especially a modeling of relationships to other players and of player immersion within the FPS game acoustic ecology. Throughout this chapter (and it is a common thread in earlier chapters), I have continually referred to the importance of the ideas discussed here to the perception of immersion within and participation in the FPS game acoustic ecology. Diegesis, as a way to explain sound-based player relationships in a multiplayer game, and immersion in the FPS game acoustic ecology, are the subjects of the following chapter.

Chapter 7

Diegesis and Immersion

Introduction

The previous chapters have dealt with a variety of aspects of sound in the FPS game as they have relevance to the hypothesis of the FPS game acoustic ecology. Thus, to name a few of these, I have discussed sound *sui generis*, relationships between sound and image and sound and event, and the meanings derived from these conjunctions, and I have described how sound can create perceptions of spaces, locations and times in the game. What generally has been missing from the research so far and what it has, in part been leading up to, through its signalling as a common thread in previous chapters, is player immersion in the acoustic ecology. Furthermore, where I have previously briefly used the terms diegetic sound and nondiegetic sound, I now fully define these terms and refine and develop them for the medium of the FPS game by establishing and defining new terminology as aids in the discussion of immersion and players' sonic relationships. Having provided taxonomies of FPS game sounds, a range of perceptual factors influencing sonic meaning and accounted for the relationships between sound and image and sound and game event, this chapter now attempts to account for the influence of sound upon player immersion and the relationships between players and, in so doing, expands the model of the FPS game acoustic ecology further. In particular, the modified model presented in the conclusion not only includes immersive and diegetic factors but also explicitly accounts for player (sonic) interaction in a multiplayer FPS game.

The various concepts of diegetic sound which I here provide (especially the concept of telediegesis) are important for the development of the hypothesis and so are a major contribution to the conceptual framework. Such concepts help to differentiate in particular the varying sounds experienced by, and, therefore, the varying relationships to the acoustic ecology of, each player in a multiplayer FPS game. They also aid in clarifying some of the relationships between players in a multiplayer game. I have previously stated that most sound in the FPS game is predicated upon player action (chapter 5) and so a discussion about the diegetics of FPS sound will help illuminate such relationships. Furthermore, diegetic FPS game sound in its various guises is intimately connected to immersion in the acoustic ecology through the establishment of such player relationships.[1]

As a further preparation for the discussion on immersion, I briefly discuss realism in FPS games and suggest that a reduced, perceptual realism may be all that is required to form the basis from which immersion can work. Where I have previously used terms such as immersion, participation and performance furnishing merely brief explanations, here I more fully explore these terms in the context of the relationships between players and between players and the soundscape in the context of the

---

[1] Although it is multiplayer games that are referred to with the use of 'players', some of these relationships also exist between player and bots in a single-player game with various exceptions (bots do not hear sound and so there is no telediegesis) which are discussed further below.

spaces (chapter 6) the player experiences and is immersed in. It is the player who is immersed in the ecology, it is the player who participates in it and it is the player whose performance helps create, sustain and transform it. Immersion in the game can be part-explained through a conceptualization of immersion in the acoustic ecology. This conceptualization is the subject of this chapter.

7.1 Diegetic sound

Sounds heard when playing an FPS game derive either from actions and events within that world or they do not. Given the principles of sonic causality, affordances and the perception of resonating spaces and paraspaces (to mention some of the ideas discussed throughout this work thus far) and the import of these principles to the hypothesis, it is therefore important to be able to state which sounds in the FPS game derive from game world actions and events and which do not. The former form part of an understanding of the game's acoustic ecology whilst the latter do not. A conceptualization of an acoustic ecology helps to explain actions and objects in the game world because it comprises sounds produced by those actions and objects. Furthermore, because the player's character and his actions belong to the game world, an acoustic ecology provides an understanding of character context. Any sounds heard while playing the game that do not inform about the game world (such as the musical score as I briefly discuss below) are not considered a part of the acoustic ecology framework. This section explores those sounds that *are* a part of that framework (in other words, diegetic sounds) and their diegetic relationship to the player.

The terms diegetic and its antonym nondiegetic are widely used in film theory to discuss and analyze cinematic sound (see Altman, 1992; Chion, 1994 for example). Nondiegetic — as in bearing no relation to the diegesis or primary narrative — is a term Chion uses in preference to extradiegetic which refers to events and actions outside the primary narrative but still containing a connection to it (2003-2006). Nondiegetic sound in film sound theory generally includes the dubbed-on music score as it is usually composed (or selected and edited) as one of the last stages of film production (this is certainly the case for most live-action films of the popular commercial cinema variety but there are exceptions to this model such as music documentaries and music videos).[2] I use the term nondiegetic in preference to extradiegetic firstly because it is the more commonly used term in film sound theory and secondly because it is the term used by game music theorists (Whalen, 2004 for example) following the lead of their film theorist colleagues.

The construction of diegetic/nondiegetic sound terminology as regards film is therefore intimately concerned with the film's narrative and the existence of a border between the story world and the real world; on the one side is diegetic sound, on the other nondiegetic sound. However, not all theorists accept that such terminology applies to all modes of film. Curtiss (1992) states that it has little or no use when discussing cartoons and prefers the terms *isomorphic* and *iconic*, the former relating to similarities between music and image (ascending scales mimicking ascending

[2] It is arguable as to whether any film musical score really has *no* connection to the primary narrative or diegesis of the film. As an example, film music is often composed to heighten emotionally the events portrayed on screen. Furthermore, as pointed out below, an isomorphic musical soundtrack (chapter 5) is intimately connected to the events portrayed on screen.

visual patterns for example) while the latter is used in the Peircian sense of a purported one-to-one relationship between sound and image (a cymbal crash when a cartoon character is hit on the head for example) (p.202). This latter methodology clearly has relevance to the study of sound in certain genres of digital games (platform games for example where 'cartoonish' sounds are used) but it also has relevance to FPS game sound as I have explained in chapter 5.

As regards the music soundtrack of the FPS game, the argument could be made that it has strong connections to the diegesis. I am not suggesting that the FPS game is designed around a music composition or compositions; I am suggesting that the music can have an effect upon the action experienced in the FPS game. In the case of popular commercial cinema, this rarely happens (some editors and directors such as Stanley Kubrick edit or edited to a musical rhythm), and I have already mentioned the use of pre-composed music in some types of cartoons in chapter 5. The music in many FPS games exists somewhere between these two poles. From a personal point of view, the music in *Quake III Arena* (id Software, 1999) can, on occasion, affect and even effect my movements and actions on screen as my character (through my controlling haptic input) moves to the beat or fires to the rhythm of the music.[3] Here, my actions are isorhythmic to the music and the music has an effect not only on my actions but, through my actions, on the diegesis and gameplay experienced by other players as I miraculously dodge obstructions and projectiles with a rhythmic grace and flair I only wish I could duplicate in reality — an animated *danse macabre* that may be likened to Curtiss' description of animators animating to the isomorphic sound of the Warner Bros. cartoons (see chapter 5).

If, however, an altered definition of diegetic sound is applied, then the musical soundtrack of the FPS game can be categorically described as nondiegetic sound. This amendment to the standard cinematic definition (sound that is from within the same world as the characters inhabit on screen (Curtiss, 1992, p.201)) is required because of the implications for the gameplay of the player's ability to kinaesthetically interact with the game world and because of the multiplicity of diegetic outcomes any one playing of the game may have (two of the major differences to film where the terms diegetic sound and nondiegetic sound have been most heavily theorized). I propose, therefore, that diegetic sound in a digital game be defined as the sound that emanates from the gameplay environment, objects and characters and that is defined by that environment, those objects and characters. Thus it includes, for example, footsteps, ambient sound, gunfire, voice radio messages (and here, third-party software that is not a part of the packaged game is also included for this purpose[4]) indications that other characters have entered or left the game and notifications of the attainment of significant gameplay targets (such as the capture of a flag). In other words, for an FPS game sound to be diegetic, it must derive from some entity of the game during play. Such a definition conceptually separates the FPS musical soundtrack from all other gameplay sounds more satisfactorily than does the cinematic definition of diegetic sound. It also closely relates diegetic sounds to the resonating spaces and paraspaces discussed in chapter 6 as these spaces are created by sounds which emanate from the environment, objects and

---

3 The character Anarki is particularly appropriate for this as she balletically zips around the arena on a hoverboard. In most FPS games, I turn the music off completely (in order to hear the diegetic sounds better) but not in this one. Here, comparisons may also be made to the juxtaposition of music and violence in the cinematic style of the more operatic film directors as I have already noted in chapter 1. Both emphasize lurid excess.

4 Although, strictly speaking, such software usually bypasses the game engine.

characters of the game world.

By contrast, nondiegetic sound refers to all sound events prior to and following gameplay; this includes a variety of interface sounds heard when the user interacts with the game or level configuration menus and any musical score. Similarly, nondiegetic sound may also be considered to be the musical score that can often be heard throughout the gameplay (and that usually has its own volume control separate to other sounds in the configuration menu). From this point on, having defined nondiegetic sounds in order to aid in the definition of diegetic sounds in the FPS game, this chapter deals solely with diegetic sound and its relevance to the acoustic ecology of the FPS game as signaled in chapter 1.

In the case of networked multiplayer FPS games, the definition of diegetic sound, while it may to some extent indicate a relational position between the player and the events of the game world, does not help to illuminate relationships between players. To further this aim, the definition of diegetic sound must be further refined. In this type of gameplay, there are different acoustic spaces (as defined in chapter 6) geographically dispersed at each player's audio interface; depending on the player's virtual proximity to other players within the game and the audio zoning circumscribed by the game engine,[5] these acoustic spaces may be individual to each player, may be shared or may overlap dynamically with other acoustic spaces as players move their characters and the game progresses. Some of the component sounds within these spaces may be heard in different contexts and with different audio processing by a variety of players simultaneously, may be private to one player (such as the breathing of the character in some FPS games) or may be private to a team (such as voice radio messages). All of these sounds are diegetic in that they are intimately concerned with the gameplay action and, in many cases, have an effect upon that action as players respond to them.

The fact that a sound may have an effect upon the action and the fact that not all sounds can be heard by all players at the same time are two of the main differences between film sound and FPS game sound and these differences are what problematizes the use of the term diegetic sound, as derived from film sound theory, for FPS games. The solution is to make a distinction between various types of diegetic sound heard during FPS gameplay and for this I propose the terms *ideodiegetic sound* and *telediegetic sound*. Ideodiegetic sound refers to all gameplay sounds that can be heard by one player; all sounds within that particular player's resonating spaces whether they derive from the player's character or from other sound sources in the vicinity. Telediegetic sounds are sounds that are ideodiegetic to one or more players but, if unheard by another player, are telediegetic for the latter player if the response to the sound by the former has consequence for the latter. Any of the audio samples in the FPS game may be used ideodiegetically or telediegetically at different audio interfaces throughout the networked multiplayer game domain — context is all important.[6]

---

[5] The game engine will, in most cases, specify when sounds may or may not be heard depending upon a range of factors including the virtual distance between the listener and the sound source and the existence of walls and other virtual sound-absorbing materials between listener and source.

[6] The reader should not forget that FPS game audio samples are not delivered over the network from the game server or from one player's computer to another player's computer (except in the case of third-party radio communications). A player's game engine receives instructions from the game server to play particular sounds dependent upon the gameplay.

Ideodiegetic sounds, therefore, are all sounds heard by one player and comprise proprioceptive sounds (such as the sounds of the character's breathing) and exteroceptive sounds (which are all other ideodiegetic sounds that the player hears whether these sounds originate from the player's actions within the game or are the sounds of other characters or game entities within the zone of hearing as defined by the game engine). Telediegetic sounds are sounds in a multiplayer game which are not heard by one player or several players but are ideodiegetic to another player or players. Because the latter player can kinaesthetically react to his ideodiegetic sounds, thereby having an impact on the sum multiplayer gameplay and its outcome, these unheard, telediegetic sounds therefore have consequence for the former player. For example, player A, guarding the team X base in a capture the flag game, hears the ideodiegetic sounds of distant battle and, using the navigational listening mode, moves to join the fray. Player B, on team Y, having not heard the aforementioned sounds, enters team X's base and discovers there is no longer anyone guarding the base and this enables him to steal team X's flag. The sounds, ideodiegetic to player A, are telediegetic to player B because player A's reaction to them has later consequence for player B. In a single-player FPS game configuration, all diegetic sounds are ideodiegetic — there are no telediegetic sounds because bots do not hear or react to sound.[7] Some ideodiegetic sounds in the multiplayer FPS game may be classed as *exodiegetic* in that, although they may be heard by the player, they do not result from that player's haptic input. Those that do, I term *kinediegetic*.

There is a phenomenological question that, at first sight, seems to negate my distinction between ideodiegetic and telediegetic sounds: *if a leaf falls in a forest and no-one is there to hear it, does it make a sound?* For telediegetic purposes, the point becomes not *does it make a sound?* but *does it have consequence?* Given the players' abilities to react to ideodiegetic sounds and therefore impact upon the gameplay, then this impact is likely to be of consequence to another player — in other words, sounds that are telediegetic to that player still have import for that player. Thus, despite the telediegetic sound being unheard by the player, it has a consequence for that player because it is intimately connected to the gameplay action as either being, in part, representative of that action or, more importantly in this case, provoking a reaction on the part of other players the effects of which eventually impact upon the initial player.

Röber and Masuch (2004) suggest, in the context of audio-only games, that sound can realize objects (and, by implication, events) in the game — if an object or action has no sound, it does not exist for the listening player. To some extent, the same can be said for sound in audio-visual FPS games but with a caveat in the case of multiplayer configurations. Sound is not alone in perceptually reifying objects and actions if they are in sight of the player; in the reification of that object or action, hearing works in conjunction with vision. In the case where objects and actions are not in view of the player, then sound alone has this important perceptual function as long as it is ideodiegetic sound.[8] If the sound cannot be heard then the object or

[7] They may react to the presence of other game objects or characters but this reaction is one based on other factors such as collision zones and vicinity that are defined by the game engine and of which sound (heard by the human player) is merely an indicator. Despite the deafness of bots, the player's reaction to ideodiegetic sounds may have an effect on bots' later actions and thus there is still some sense of telediegesis but, in this case, the effect moves solely from player to bot.

[8] There are many other factors, particularly cognitive, at play here which the hearing of a sound triggers. However, the starting point is the sound of the unseen source.

action does not exist at the present time for the player. The exception to this is the case of telediegetic sound in a multiplayer configuration. Although it can be said, at the precise moment in time that the telediegetic sounds are heard ideodiegetically by another player or players and that the objects and actions associated with the sounds do not exist for those who cannot hear the sounds, the argument can be made that these telediegetic sounds of objects and actions later reify at least their implications if not the objects and actions themselves.[9] Thus, if the objects and actions do not exist momentarily for one player because they cannot be heard ideodiegetically, potential responses by other players to these reified objects and actions, that have later consequences for the first player, mean that these objects and actions have a persistence of hearing.

Following film sound theory, then, FPS game sound may be defined as diegetic or nondiegetic. However, because players may kinaesthetically interact with FPS game sound and because players in a multiplayer FPS game will hear not only different combinations of some sounds but, in many cases, different sounds altogether, the definition of deigetic sound in the FPS game must be expanded to suit this different, non-cinematic medium. An FPS game diegetic sound may, therefore, be defined as ideodiegetic or telediegetic (except in single-player configurations where they are solely ideodiegetic). Ideodiegetic sounds may be classed as exodiegetic or kinediegetic. These definitions are a matter of provenance, context and utility at any one point of time in the game. All diegetic sounds have relevance to the FPS game's acoustic ecology because they inform about events and objects within the game world which are expressed through sound. Nondiegetic sounds have no relevance *per se* to the FPS game's acoustic ecology as they do not inform about events and objects in the game world.[10] The concept of telediegesis proves to be a particularly useful one when conceptualizing the multiplayer FPS game acoustic ecology from the point of view of dramatic performance and autopoiesis (see chapter 9).

7.2 Perceptual realism

Many FPS game designers attempt emulations (or at least high-level simulations) of reality in their games. As mentioned in chapter 6, sound in the FPS game is (currently) the only way of inserting artefacts from reality into the virtuality of the game. For example, recordings of real-world weapons which are triggered whenever the player shoots a gun within the game. Ward (2002) has the view that there are elements of digital games which tend towards mimetic realism (as close a mimicking of reality as possible) and aspects that tend towards abstract realism (pp.124—127). I would suggest that, in general, the use of sound for realism purposes falls somewhere between the two poles. Specifically, though, there are some sounds which are mimetic in their representation of realism (Doppler effects, recordings of real weapons, for example[11]) and others that are more abstract in such representation (the caricature sounds identified in chapter 4). This practice is the case with 'realism' games and mods such as *Urban Terror* (Silicon Ice, 2005). In this

[9] This assumes that the telediegetically-hearing player responds to those sounds with gameplay input or autopoietic perturbations having a ripple effect such that there is later consequence for the unhearing player.

[10] In cases where the musical soundtrack volume is so high as to override the diegetic sounds, then such nondiegetic sounds would have a destructive impact on the acoustic ecology.

[11] Here, there are parallels to film sound FX such as the shotgun in *Terminator 2* (Cameron, 1991) which is actually produced from a recording of two cannons (chapter 5).

type of FPS game, the majority of audio samples used may be ascribed such a mimetic realism and this is often combined (where the game engine allows it) with real-time processing of audio samples for depth or reverberation cues depending on the materials and spaces of the game in an effort to emulate soundscapes found in the real world. However, several authors suggest that, for the purposes of immersion, a reduced realism may be all that is required. A realism that is not entirely real. This section investigates this possibility and the implications of this for immersion in the FPS game acoustic ecology.

In order to provide a sense of realism, it is not necessary to provide first-order reproductions of materials or phenomena from reality within the game environment. First-order reproductions of materials and phenomena refer, in the case of images, to photographs of game object correlates taken from reality and, in the case of sound, to audio samples (taken from reality) of the objects and actions of the game. Chion, writing about film sound, states that what is necessary to provide a sense of realism may in fact be quite opposed to any state of reality and is often the subject of convention. There are sound conventions and "specific codes of realism" that produce anything but the sound that exists in reality but rather provide "the impression of realism [and these conventions become] our reference for reality itself" (Chion, 1994, p.108). This comment about the nature of reality and the role played by sound in the creation of realism is supported by other writers on film sound and by those writing about sound design and use in other audiovisual media. Lastra (2000) states that: "Decades of tin-sheet thunder and coconut shell hooves prove [...] that fidelity to source is not a *property* of film sound, but an *effect* of synchronization" (p.147). He views the recording and compilation of stock sound effects to be later used for dubbing purposes as symptomatic of the move from the fidelity of the original to a constructed representation of reality where "[s]ound space, no longer theoretically defined by the passive perceptions of a securely located (and physically real observer), is now shot through with the hierarchies of "dramatic" relationships" (Lastra, 2000, p.207). In other words, the meaning and verisimilitude of the sound is more important than fidelity to the original; more important than providing and using an authentic sound.

Fencott (1999) states that "the mental constructions that people build from stimuli are more important than the stimuli themselves" when discussing presence in virtual environments. This is an argument that sensory fidelity to the original recording, as part of the urge to make the sounds in a digital game as 'realistic' as possible, is perhaps less important than perceptual verisimilitude. Laurel (1993) provides a definition of a virtual object within a computer desktop or virtual environment that is "one that has no real-world equivalent, but the persuasiveness of its representation allows us to respond to it *as if* it were real" (p.8). In this sense, the *reality* of *virtual reality* is nothing more than a perceptual realism that, in the 'persuasiveness of its representation' allows the user or player to perceive it as if it were real. Back and Des (1996) discuss methods of processing sound, a recording of the original sound, such that it becomes a caricature of the original sound that is more persuasive in its representation of the sound source than a more authentic sound would be. The principles of such caricatures may be related to ideas of sonic metaphor and symbolism and provide some support for the use in FPS game sound design of Gaver's 'caricature' algorithms described in chapter 5. Realism in the FPS game, therefore, works through a system of perceptual realism based on verisimilitude, a realism of theme in which there is "plausibility of characterization, circumstance and action" (Corner, 1992, p.100) as opposed to a realism that strives for resemblance to real-world objects with, for images, photography being the yardstick (Darley, 2000, pp.16—17). In all cases, plausibility, or believability, is supported by convention and consistency in the use of sound as discussed in chapter 4.

Chapter 5 stated that the sound of the FPS game is the original as it is virtually indexical to and causally associated with the in-game events. Furthermore, sound (as I explain in chapter 6) is the one element of the game world that is 3-dimensional and that, in its presentation of sensation as primary experience, is real in terms of one space it fashions (the real resonating space), as opposed to entirely virtual. These two points, at first sight, appear to contradict the previous statements by Lastra and Laurel and the conclusions drawn from them above. However, this is not the case at all. Sound is capable of operating on several levels especially in the world of the FPS game. The images on screen can only be a part of a virtual environment as they are a 2-dimensional representation of 3-dimensionality. Sound, though, exists and operates both in reality and in virtuality; it has a real volume and dimensionality that is a real 3-dimensional representation of the 2-dimensional representation of the 3-dimensional world of the game.[12] Yet, in managing this feat, sound is also illusory not solely because if refers to a virtual resonating space but also because it may make use of caricature and convention rather than (necessarily) authentic sound to represent the variety of paraspaces of the game. It is the player's task to correlate reality and virtuality through a form of 3-dimensional synchresis — the meta-synchresis I referred to in chapter 6.

It may well be a mistake, then, to strive too hard for realism, an emulation of reality, when designing the sounds of an FPS game. Perhaps what is more important is to create a reduced realism, operating at the level of convention, consistency and plausibility in its efforts to persuade the player of its reality, and which is coupled with the predication of (most) sounds upon player action (chapter 5) and all the other codes of realism extant in the game (such as simulations of gravity and social rules, for example). A desire for realism should, perhaps be balanced with the recognition of the very visceral qualities for which sound may be used especially in FPS games — the SPAS shotgun of *Urban Terror* or the shore-based artillery of *Battlefield 1942* (Digital Illusions, 2002), with the right audio hardware[13] have an immediate, physical impact as a result of enhanced bass frequencies coupled with a high amplitude.[14] This reduced realism then, may be balanced by sensory enhancement (increasing sensory immersion, therefore) such that the resulting soundscape falls between van Leeuwen's (1999) naturalistic and sensory coding orientations for sound (pp.177—182), between an accurate representation of what one would hear were one physically present in the game world and an increased emotive impact at the expense of natural realism. Thus, as far as realism is concerned, a reduced realism, a perceptual (rather than naturalistic) realism may be all that is required as a foundation for player immersion within the FPS game acoustic ecology.

---

12 It is possible to add further layers to this particular conceptual onion or matryoshka doll by suggesting that sound will always carry artefacts of the user space and equipment that is inhabited and used by the user in order to partake in the game and that this adds, or certainly mixes in, another real space with any other real and illusory spaces already contained in the sound. One can only admire the abilities of the auditory system in being able to sift through and strain this miscegeny for game-pertinent information.

13 That is, having a large dynamic range.

14 One of the reasons why *Terminator 2* uses a recording of cannons for the shotgun sound.

7.3 Immersion

One of the aims of the FPS game is to perceptually immerse the player within the game environment and the first stage of immersion in the game may be said to occur upon the reading of the game's back-of-the-box marketing. However, during the playing of the game, if players are to participate in the construction of an acoustic ecology and are to be able to contexualize themselves within the resonating spaces and paraspaces discussed in chapter 6, immersive factors in the FPS game acoustic ecology are an important component to consider. Furthermore, an understanding of how this acoustic ecology aids in creating this perceptual immersion helps to explain some of the relationships between players and between players and the game engine which are founded upon sound. Because elements of realism in the FPS game (in this case a perceptual realism) are closely related to perceptions of immersion, I also explore aspects of FPS game sound which contribute to such realism. This section, therefore, investigates ideas of immersion and perceptual realism in digital games as they contribute to an understanding of the concept of the FPS game acoustic ecology. It should be noted that the immersive functions of FPS game sounds will be affected by the player's knowledge and experience (as noted in chapter 4) and this is expanded upon at various points during this discussion.

The goal of the FPS designer is to make the player believe that he is within the game environment, that he is the character whose hands he sees before him. Back stories and promotional material always address the player in the second person singular and situate the player within the world postulated by the game: "[Y]ou are U.S. Army Ranger B.J. Blazkowicz [...] You are about to embark on a journey deep into the heart of the Third Reich" (Gray Matter Studios & id Software, 2001), "you lunge onto a stage of harrowing landscapes and veiled abysses" (id Software, 1999), "[y]ou are a marine [...] Only you stand between Hell and Earth" (id Software, 2004). This is supported by advances in computer technology, both hardware and software, where "incredible graphics, and revolutionary technology combine to draw you into the most frightening and gripping first person gaming experience ever created" (id Software, 2004).[15] It is my argument that perceptual realism, and therefore immersion, can be brought about and greatly enhanced through the judicious use of sound and if, as previously mentioned in chapter 5, sound is perceived prior to vision, it may be that sound is the first line of attack in the creation of a perception of realism and immersion in 3-dimensional virtual environments such as those found in FPS games. As Anderson (1996) states writing about film, "sound is seventy percent of the illusion of reality [perceptual realism] in a motion picture" (p.80). Although I am not as willing as Anderson to provide such a precise figure, the arguments below support my contention that sound is of great importance, if not the greatest importance, in creating a perceptual realism in FPS games which leads to immersion.

The use of the terms 'players' and 'characters' above points to an important difference between FPS games and films. A character and player in the FPS game are, in so far as the perception of sound is concerned, one and the same due to the first-person (visual and sonic) perspective.[16] Where film sound practice and theory has moved from the notion of impossible auditors through to external and internal

[15] This form of direct address occurs not only in FPS games as Burn and Parker (2003, p.45) note.

[16] In fact, a player almost never sees his character other than as a pair of hands or a single hand, when the FPS game features mirrors within the game environment, in overhead shots of the player's dead character while waiting to respawn or other uses of third person perspective as detailed in chapter 1.

auditors, there is no such distinction in FPS games where the immersive, 3-dimensional nature of the game posits the player as first-person auditor (chapter 5). If the diegetic graphical locus of the film exists solely for the characters on screen, then the diegetic sonic world of the FPS game extends from the screen to physically encapsulate the player too in the acoustic ecology's real resonating space (chapter 6). This is particularly the case where the player is using headphones as these serve as an extension to the player's proprioceptive auditory system greatly attenuating, and in some cases entirely blocking out, sounds external to the game world such that, for example, the sounds of the character breathing become the sounds of the player breathing.[17] Thus, FPS game diegetic sounds extend the game environment from the flat, 2-dimensional screen to the 3-dimensionality of the external world. The player's proprioceptive sounds are replaced by the character's proprioceptive sounds, all other game world sounds (both kinediegetic and exodiegetic and with appropriate sensory immersion) envelop the player as part of the game's real resonating space.[18] These sounds form part of not only the real resonating space but also the virtual resonating space of the game and thus help to immerse the player, both physically and mentally, in the FPS game acoustic ecology.

In his book *Visual Digital Culture*, Darley states that 1995 saw the release of "the first feature-length computer synthesized film *Toy Story*" (p.20) in a chapter where the impression is given that realism and immersion in digital media are mainly due to, if not solely, the power of graphics. This statement is wrong. *Toy Story* (Lasseter, 1995) was not 'the first feature-length computer synthesized film'; parts of it are 'computer synthesized', including all the graphics, yet no-one would argue that the voices in the songs or the voices of the actors are synthesized. (It is interesting to note that while computer graphics can produce stunningly realistic animations (witness the recent work of Weta Workshops), voice synthesis lags far behind.)

However, Darley's statement does serve to indicate the lack of attention given to sound not just as something to listen to in the context of digital media but as a vital component in the creation of a sense of perceptual realism and immersion in the 3-dimensional media such as FPS games. The method of production of graphics and sound that is found in films like *Toy Story* has similarities to FPS games which, while the graphics may be synthesized on a computer, rarely use synthesized sounds but instead tend to the use of audio samples of recordings of real-world correlates of objects found in the game (some of which, admittedly, may be recordings of synthetic sounds) or foley sounds. It is pertinent, therefore, to consider what role such audio samples play for the player in creating a sense of immersion in FPS games. It is my contention that, while image may provide some of the illusion of immersion within the 2-dimensional visual environment of the game, sound, with its 3-dimensionality, its real and virtual resonating spaces, provides for physical and mental immersion within the acoustic ecology of the FPS game and, in so doing, enables and completes the game's immersive illusion partially initiated by the image.

Ermi and Mäyrä, paraphrasing Pine and Gilmore (1999), state that "immersion means becoming physically or virtually a part of the experience itself" (Ermi & Mäyrä, 2005). As part of their 'four realms of experience', they define games as an escapist

[17] A form of sensory immersion which is discussed below.

[18] Aided by their technical similarities to the game character's proprioceptive sounds, they are co-opted by the player in like manner.

experience because they include both immersion and active participation.[19] The authors use this as a starting point to analyze immersion in thirteen digital games where they posit three types of immersion: *sensory immersion* where sensory stimuli from the game (auditory and visual) override sensory stimuli from reality; *challenge-based immersion* requiring motor and mental skills and *imaginative immersion* where players identify with the game's story and characters.

From the results of their surveys, it is interesting to note that *Half-Life 2* (Valve Software, 2004), one of two FPS games in the survey, not only scored highest in terms of greatest sensory immersion but also had the highest overall immersion score (including the second highest imaginative immersion rating) (Ermi & Mäyrä, 2005). Despite apparent flaws in the methodology,[20] their classification of different types of immersive experiences proves to be a useful starting point for describing the immersive function of sound in the FPS game acoustic ecology.

All sound in the game has the ability to provide Ermi and Mäyrä's sensory immersion but, because this is sensory rather than perceptual, this level of this type of immersion is dependent upon the type of audio interface the player uses, the relative volumes and relative frequency bandwidths (and overlap) between the game environment's sounds and the player environment's sounds. However, it is almost certainly the case that sensory immersion is increased through the wearing of headphones (as most FPS players do) because these serve to block out sounds from the user environment. Where, according to Ermi and Mäyrä, immersion includes becoming *physically* a part of the experience, this is achieved, in the case of the FPS game acoustic ecology, through the process of the player becoming physically immersed in the real resonating space (chapter 6) and so sensory immersion is a physical immersion.

A range of FPS game sounds can provide a challenge-based immersive function requiring the use of kinetic and mental skills. The sound of grenades being launched from the grenade launcher in *Quake III Arena* provides an apt example. If the target is close enough to the firer, the sound of the firing will be heard; if the target is close enough to the falling grenades, the sound of grenades falling, like clattering metallic hail, will be heard; in some cases, both sounds will be heard depending on the relative positions of the attacking player and the target and on the accuracy of the firing. Mental skills are required on the part of the target. Firstly, and once the sounds have been causally assessed as grenade sounds, a threat assessment must be carried out — in teamplay configuration, the server variable *teamDamage* may be off in which case the grenades are no threat if fired by a team member. In all other cases, the grenades are a potential threat. Secondly, and based on prior experience of the sounds, the relative amplitude of the sound of grenades falling can be used to assess the danger — relatively quiet sounds mean the grenades are too far away to cause any damage. Thirdly, the direction of attack must be ascertained based on the localization affordances offered by the sound of grenades skittering around the target compared to the localization affordances, if available, of the sound of the grenade

---

[19] Defining absorption as "directing attention to an experience that is brought to mind", the other three realms of experience are: *entertainment* (absorption and passive participation); *educational* (absorption and active participation) and *aesthetic* (immersion and passive participation).

[20] There is little description of the methods used and, where one of the survey ranking statements about sensory immersion is: "The sounds of the game overran the other sounds from the environment", there is no description of sounds in the environment, the game or environment sound pressure level or the audio equipment used — headphones or loudspeakers — among many questions which might have been asked.

launcher (possibly coupled with a visual sight of the attacker). All of this is used to plot a course of action which is where kinetic or motor skills and mental skills come into play. Upon assessing the direction of fire and that the grenades are a threat, the player needs to avoid their imminent explosions with a combination of kinetic and mental dexterity. This will involve decisions and actions concerning running (away from or towards the attacker), dodging or fighting other players in the meantime, taking cover behind obstacles, using various items if available (such as jump pads and teleporters), weighing up the pros and cons of the damage incurred by dropping from a height in an effort to escape or one of many other possible scenarios offered at that particular point in time and dependent upon the gameplay circumstances.

Any diegetic sound can provide challenge-based immersion but some sounds (such as the grenade launcher and grenades described above) have the potential to provide more of this function than other sounds and much of this depends on context and game experience. At first playing of a game, it can be said that all diegetic sound provides a mental challenge even if it is solely *What's that sound? I wonder if it's important?* At subsequent playings of the game, the player will have categorized the sounds into those presenting a challenge and those which present no challenge being simply part of the auditory setting, that is, environment sound.[21] Likewise, audio beacons primarily require navigational listening on the part of the player as he navigates (sometimes in quite complex mazes) towards the source of the sound. Once the route has been learned, once a mental map has been constructed, there is less of a challenge afforded by this sound. Thus, some sounds retain a high degree of challenge throughout multiple playings of the game while others have their challenge lessened or negated once the riddle has been solved and sounds have undergone the player-initiated process of categorization. Typically, those that retain some challenge-based immersive function are those whose context may change as opposed to those (such as environment sounds) whose context remains static.

Sounds affording imaginative immersion help the player identify with the characters and action in the FPS game. Some of these sounds may be described as cues or sureties (see below) because they are expected within the game environment. Therefore, if a character dropping from a height causes a heavier thud than, for example, the same character's footsteps when running, this is a mundane cue because, if virtual gravity functions within the game environment in a similar manner to which it functions in reality, this louder, more massive sound is to be expected. Similarly, a variety of causally attributable sounds within the game are also cues and this is particularly the case where the sound results from some action, such as the firing of a weapon, on the part of the player. The class of proprioceptive sounds, like the character's breathing, are especially potent immersory cues particularly if the game engine allows for a change in breathing rate following the speed and exertions of the character.

Carr (2006), summarizing definitions of immersion as provided by other writers, states that there are two categories of immersion: "[P]erceptual immersion, which occurs when an experience monopolizes the senses of the participant, and psychological immersion, which involves the participant becoming engrossed through their imaginative or mental absorption" (p.69). Although the use of the term

---

21 By now, the player will have had some experience and training in the game leading to a hierarchical categorization of affordances which in turn affect the mode of listening (chapter 4).

'perceptual immersion' is confusing as it refers solely to sensory systems,[22] Carr's description of these two categories of immersion bears strong similarities to Ermi and Mäyrä's sensory immersion and challenge-based and mental immersion. Both categories of Carr's immersion are usually at play in all digital games but the differences in affordances offered by different genres of digital games will tilt the balance in favour of one or the other. FPS games, I would suggest, operate more (but not solely) at the level of visceral, sensory immersion compared to Role Playing Games (RPG) games, for example, which, with their strongly narrative and socially interactive bent, accomplish any immersion mainly by psychological means. These categories are also similar in many respects to McMahan's systems of perceptual realism and social realism discussed in section 7.2.

Fencott defines two *details* of virtual environments that help create the mental state that he terms presence. *Cues* are mundane and predictable and include things such as expected objects (doors on houses, trees in a park), elements of scale and distance and, importantly, causal sound (chapter 5). *Surprises* are the second detail and work to maintain interest, as opposed to cues which are accepted as given. They might include a video playing in a picture frame or sounds having no discernible cause. It is important to note that cues function through user experience and knowledge (gained either through playing the game or gained culturally and external to the game) and that surprises, being surprises, require a certain ignorance (certainly on first hearing the sound) on the part of the player; in cognitive terms, the cue is already categorized while the surprise requires categorization.[23] Nevertheless, surprises, like cues, must be derivable from the logic of the environment if the sense of presence is not to be shattered; both details complement and support each other and the virtual environment "by seeking to both establish fidelity and catch and retain the attention of the visitor" (Fencott, 1999).

Such details of the environment may also be discussed in terms of affordances as they are not only opportunities to *do* things in the virtual environment (for example to open a door to enter the house in the game) but also to *perceive* things — in this case, the opportunity of perceiving the virtual environment itself. To provide an example, the sound of birds in a forest is a cue as it is mundane and expected (in both reality and virtual forests); when it suddenly stops, this acts as a surprise as our natural inclination would be to wonder what caused all the birds to simultaneously stop singing. The affordances (chapter 4) offered by this sound (when part of a virtual world or FPS game) are the opportunity to believe that the player is amidst a forest with birds singing and, when there is a sudden lack of the sound, the opportunity to be aware of possible danger. If sonic affordances may be presumed to be fundamental to the provision of sonic cues and surprises then it is likely to be the case that the more affordances a virtual environment has, the greater the sense of presence or immersion in the FPS acoustic ecology.

This system of cues and surprises (and as further developed by McMahan below) may be paralleled by Malaby's (2006) description of a digital game as a system of multiple contingencies, contrived and calibrated, leading to both predictable and unpredictable outcomes (p.9). In this sense, the interpretation of aural cues and surprises is (on one level and as a simple example) contingent upon user experience

---

22 Strictly speaking, this is actually *sensory immersion*.

23 See comments on causality and familiarity in chapter 5.

of such sounds or not. These are contrived and calibrated by the sound designers and game designers in an attempt to produce the desired mix of both predictable and unpredictable outcomes of player responses that provide the patterned, yet differing, gameplay experience each time the game is played.[24] This contrived and calibrated mix must be carefully judged. As Steve Johnson states: "If games are too hard they're boring, and if they're too easy they're boring, but if they're right in the zone they're addictive" (quoted in Wasik, 2006, p.33). Although this statement is applied to game elements in general, it may also specifically be applied to the affordances of the acoustic ecology; such affordances are not accidental but designed.[25]

McMahan (2003) gives three conditions for an immersive experience: "the user's expectations of the game or environment must match the environment's conventions fairly closely [...] the user's actions must have a non-trivial impact on the environment [...] the conventions of the world must be consistent" (pp.68—69). She develops Fencott's cue and surprise details by describing cues as *sureties*, adding a third detail (*shocks* which are anything detracting from the sense of immersion and are often due to poor design — presumably details of this type should *not* form part of a virtual environment — or are external stimuli intruding upon the game world) and provides three categories of surprises. These are: *attractors* which tempt the user to do something; *connectors* helping the user to orientate himself and *retainers* causing the player to linger in and enjoy parts of the environment. Sounds with such functions are ideodiegetic sounds. For McMahan, as realism is one of the defining elements of immersion, the details of sureties and surprises are part of the system of perceptual realism that works together with *social realism* to create a sense of immersion in the game.

Sounds that function as attractors, tempting the player into some action, will be, in the FPS game, sounds such as the enemy's fire. The example of the sound of the grenade launcher given above is a good instance of this type of sound compelling the target player into some type of response. Other such attractor sounds may be certain announcements or, in particular, teamplay sounds such as a sound or a voice indicating that a team flag has been taken in capture the flag configuration. In this case, the attractor functions to prompt either defensive or attacking play around the flag carrier depending on which team a player is a part of.[26]

A connector sound is one which helps the player orientate himself within the illusory 3-dimensional spaces represented on the screen. The ability to localize sound and to discriminate depth is important here, less so when a sound serves to indicate that a player is already within a certain location within the game environment, more so when the player wishes to turn to or move towards the sound source. Any ideodiegetic sound, with the exception of global game status sounds and team voice radio messages (usually location-less in their sonic quality), can serve as a connector and connector sounds can be utilized for navigational listening. Although it may be that McMahan's intended use of connectors as orientators refers solely to orientation in space, FPS game sound can also serve as a connector orientating oneself in time,

---

[24] Parallels and similarities may also be found with numerous other methods of describing the same gameplay experience; gameplay as dramatic performance (see chapter 9) is a contrived mix of cues and surprises resulting in a variety of both expected and unexpected outcomes.

[25] Designed, it is to be hoped, with some thought but I accept that, in some cases, that thought may be cursory and based solely on an acceptance of popular commercial cinema sound conventions.

[26] Although, of course, the player can simply ignore this — an attractor merely issues an invitation.

whether that is time in terms of the length of gameplay or as in a specific point in time.

A sound acting as a retainer causes the player to stay in a particular locale in the game environment. In this sense, a retaining sound is also an attractor (because it tempts the player to do something, that is, to stay in the one place) and is also a connector (as it helps to orientate the player within that locale). Many FPS games, particularly of the run and gun subgenre, have few sounds whose primary function is to act as a retainer unless the whole set of game sounds is treated as part of the attraction of buying a game and playing it. Sounds may be lingered upon and pleasure derived from them at the first playing of a game, but this particular type of engagement with sound is not conducive to a high survival rate in the more frenetic FPS games.[27]

Sonic sureties and surprises are not just exodiegetic sounds from the game environment or from other characters but may also be kinediegetic sounds produced by the player's character. In this case, such sounds are typically sureties being expected and consistent details conforming to the game world's logic and conventions. Thus, when I run I expect to hear footsteps, when I fire my shotgun I expect to hear a loud blast. The causality and indexicality to player action of such sounds reinforces the immersive nature of these sureties. Furthermore, the player's production of such sounds has an ideodiegetically 'non-trivial impact' upon the acoustic ecology as shown in the two diagrams below.

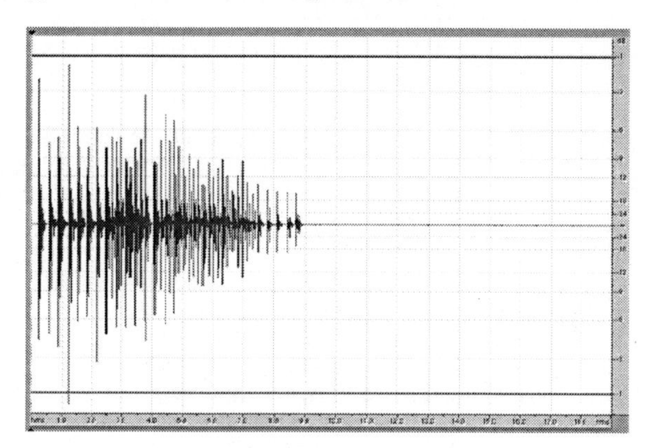

Figure 7.1. The sound heard by an inactive FPS player.

27 Although I am unaware of any such use in a commercial FPS game, I did once design a *Quake III Arena* level in which the player was invited to linger in a chamber by dint of switches and pressure pads on the walls which the player could use to activate various sounds; if the player lingered too long, the ceiling would descend upon her with force. Because this type of sound use was alien to the gameplay ethos of *Quake III Arena* this level never made it beyond the initial proof of concept.

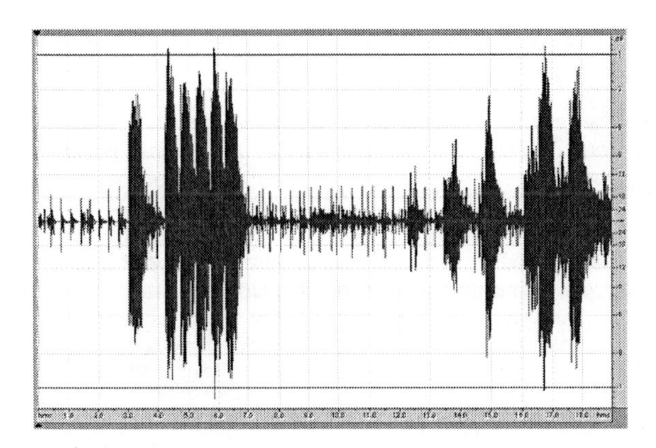

Figure 7.2. The sound heard by an active FPS player.

Both of the soundscapes shown in *Figure 7.1* and *Figure 7.2* show the sound heard by an FPS player during the first 19 seconds of an eight-player capture the flag game in *Urban Terror* on the *Abbey* level. In the first, the player is doing nothing and the only sounds heard are the soft twittering of birds and the receding footsteps of teammates as they move towards the game action. In the second, the player is running around and firing weapons. The difference between these two soundscapes demonstrates that the intervention of an active player has a great effect upon the acoustic ecology experienced by that player thereby fulfilling one of McMahan's three conditions for an immersive experience.

There are, therefore, several routes to the analysis of the immersive functions of sound in the FPS game and these provide new terminologies for the conceptual framework necessary to the hypothesis. It is important to note that immersion in the FPS game as described above is not a real physical immersion (with the exception of sensory immersion in the real resonating space of the acoustic ecology) but rather a perceptual immersion that is aided by a system of perceptual realism in which sound plays a part. Sounds may have a sensory or imaginative immersive function, or may provide an immersive function through the provision of challenges. Such immersive functions may change over time as players experience the game and thus respond differently to more familiar sounds. Sounds may also be immersively categorized as perceptual sureties, shocks or surprises and this last class consists of sounds that are connectors, retainers or attractors. Immersion in the acoustic ecology is aided by the player himself being able to have a 'non-trivial impact' on that ecology.

7.4 Conclusion

The terms diegetic and nondiegetic as they relate to sound have been theorized in Film Theory and have been applied without change to sound use in digital games. Thus, sound in the digital game is either diegetic or nondiegetic for writers such as Whalen. In this chapter, it has been shown that a more subtle distinction is able to be made in the case of diegetic sound, certainly in the case of FPS games, and that this distinction demonstrates that players in a multiplayer FPS game have different relationships to sound depending upon diegetic context. To explain such distinctions,

I suggest that diegetic sound may be subdivided into ideodiegetic (comprising exodiegetic and kinediegetic sounds) or telediegetic sound and these terms form a part of the conceptual framework that has been built up throughout this work.

Immersion and participation, as important aspects of conceptualizing the FPS game acoustic ecology, proceed on the basis of the player experience of sound and the player's ability to respond in kind, and is founded upon a system of perceptual realism which need not necessarily be an emulation of reality. Ideodiegetic sounds contribute to sensory immersion, challenge-based immersion and imaginative immersion and may act as perceptual sureties, shocks or surprises. Sound functioning as a perceptual surprise may be classed as a connector, attractor or retainer and these, combined with a logical and consistent use of sound in the game and an identification with the character's proprioceptive sounds, contribute to a system of perceptual realism. This forms one of the keys to immersion within the FPS game acoustic ecology thereby leading to immersion within the game world. It has also been demonstrated that, by participating through contributing sounds to the acoustic ecology, the FPS player is able to have a 'non-trivial impact' upon the acoustic ecology. This form of creative, sonic interaction is, likewise, one of the keys to immersion in the FPS game acoustic ecology.

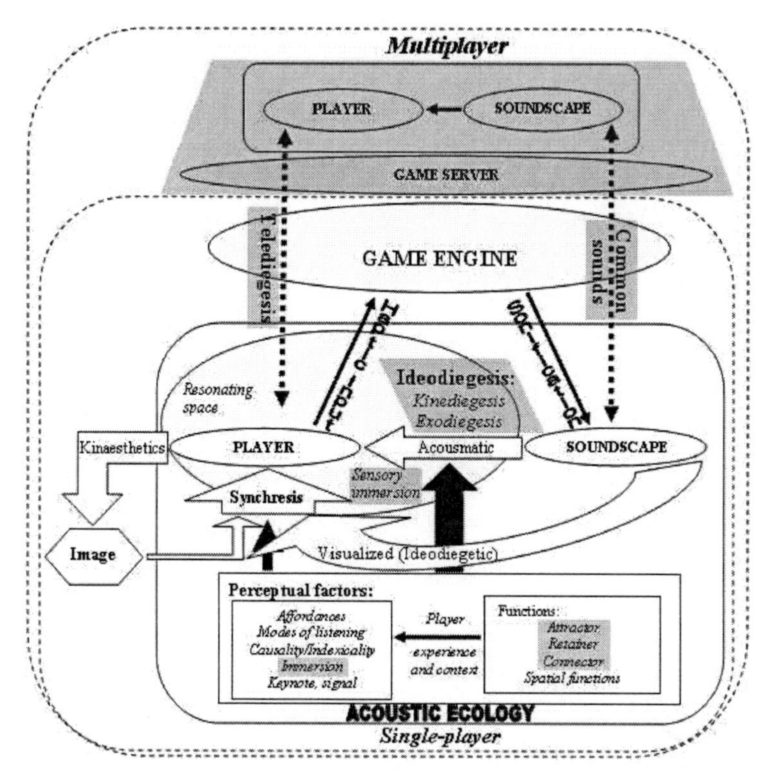

Figure 7.3. The model of the FPS game acoustic ecology including immersive and diegetic factors and detailing sonic relationships between players in a multiplayer game.

Figure 7.3 is an expansion of the conceptualization of the FPS game acoustic ecology, developed from chapter 1, and which now includes the main points raised in this chapter. The immersive functions of sound (attraction, retention and connection) are mediated by player experience and context before affecting perceptual factors influencing the meaning of visualized and acousmatic sounds. These latter classes of sound are described as ideodiegetic (kinediegetic or exodiegetic). Whereas challenge-based immersion and imaginative immersion due to sound are classed as perceptual factors influencing meaning, sensory immersion is a real, physical immersion that derives from the enveloping nature of real, resonating space. Importantly, though, the model now accounts for relationships between players in a multiplayer game through the action of telediegesis and by indicating that there are common sounds between each player's soundscape.[28] Of significance here is the opportunity, provided by the multiplayer model, to propose that, in a multiplayer FPS game, there is not just one acoustic ecology but several, one for each player. Furthermore, I present the proposition that all these acoustic ecologies are contained within a virtual acoustic ecology. Such a notion is the substance of chapter 9. Prior to this, though, a bridging chapter is provided which summarizes the conceptual framework developed thus far and illustrates it with a specific FPS game.

[28] The representation of the second player shown in the model is a simplified version of that of the first player for clarity's sake. Additionally, the game server, handling all communications between each player's game engine, does not broadcast common sounds between each player's soundscape but, rather, issues instructions to each player's game engine to play such sounds.

Chapter 8

The Conceptual Framework

Introduction

This chapter represents a shift in direction from the previous chapters and serves as a bridge, between these earlier chapters and the penultimate chapter. Whereas chapters 2 through 7 comprise multidisciplinary research with the aim of developing a conceptual framework to be used for the hypothesis of the FPS game acoustic ecology, this chapter presents a summation of that framework in preparation for its use in the exposition of the hypothesis in chapter 9. Throughout, various aspects of the conceptual framework are illustrated with examples from the FPS game *Urban Terror* (Silicon Ice, 2005). This is a typical FPS run and gun multiplayer game and one which I am quite familiar with both as player and game server administrator (chapter 1). Although its use of sound is not as sophisticated as some later FPS games (for example, it has no real-time processing of audio samples), it is still capable of offering a relatively complex sonic experience and provides a more realist approach to sound use than, for example, more fantastical games such as *Quake III Arena* (id Software, 1999). As such, it is here judged to be a good exemplar of the subgenre (and counterpoint to *Quake III Arena*) and thus suitable for the purposes of the illustration of the conceptual framework. A fuller description of this game and a summary of the audio samples it uses may be found in Appendix 1.

The conceptual framework developed throughout this work as an aid to explaining the FPS game acoustic ecology may be divided into two related parts. These are the conceptual language and the taxonomy of FPS game sound. The distinction between the two is that the language is seen as a tool for a discussion of FPS game sound using terms that can be generally applied to all sounds in the game whereas the taxonomy is a classification of individual sounds in the game that attempts to explain their roles and functions within the acoustic ecology and the wider context of the FPS game world. Some of the terminology which forms a part of the conceptual framework is derived from the variety of theoretical models discussed in preceding chapters (such as kinaesthetics, affordances, modes of listening, auditory icons, diegetic sound, sonification, causality, indexicality, soundscapes, immersion theories, to name a few). These (with the exception of kinaesthetics because it already exists in digital game sound theory) have been adapted to the medium of the FPS game either by pointing out the significant differences (between the medium in which the terminology was originally used and that of the FPS game) and adjusting accordingly or by extending the theoretical model to include new terminology where existing terminology proves insufficient. An example would be the concept of diegetic sound to which have been added the notions of ideodiegetic sound (kinediegetic or exodiegetic) and telediegetic sound. In other cases, and in order to explain more unusual or even unique concepts of the FPS game acoustic ecology, new theoretical concepts have been developed. Such is the case, for example, with the provision of terminology for the spatializing functions of sound (choraplast, topoplast, aionoplast and chronoplast).

Although the conceptual framework may serve the purpose of aiding the classification of the distinctive features of the FPS run and gun game as a subgenre,[1] this is not its primary intent. The primary intent here is to provide a tool for the explication of the FPS game acoustic ecology and this chapter, therefore, draws together all the elements of this framework from their scattered locations in the preceding chapters to form a unified conceptual framework for the analysis of FPS game sound in the general case and in the specific case. Identified taxonomies will be extracted and comparisons between them made with the aim of reducing taxonomic redundancy and the result will be the final diagrammatic model of the FPS game acoustic ecology.

The conceptual framework and examples given here refer to the multiplayer FPS game, that is, a networked FPS game in which there is more than one human player. Computer-generated bots do not (yet) respond to sound but, in tracking down or evading player-controlled characters, make use of game code variables which change according to that character's position and actions. Although there are many elements of the framework which may be applied to the acoustic ecology of a single-player game, it makes more sense to present the case for multiplayer games from which, by ignoring those elements of the theory which explain player relationships, a case for the acoustic ecology of the FPS single-player game may be later extracted.

8.1 The language

The conceptual language is a set of terminologies with which to describe and analyze the sounds of the FPS game in a general sense and with which the more specific taxonomy may be integrated. This description and analysis aims to elucidate the role of sound in the FPS game and the ways in which it relates to other elements of the game, such as images and players, and this then serves as the foundation for the following treatment of the FPS game acoustic ecology in chapter 9. As before, the chapter concerns itself with diegetic sound only.

All sound in the FPS game is designed and placed in the game for one or, more usually, several purposes. Therefore, it may be said that all sound in the FPS game offers one or more affordances (see chapter 4). The astute and experienced player is able to efficiently prioritize and, therefore, utilize or ignore these affordances. The less-experienced player, or those whose socio-cultural experience may preclude a quick grasp of the variety of affordances on offer, is likely to be less adept at navigating and understanding the signs of the FPS acoustic ecology, but this system may be learned.

The understanding of sound in the FPS game world is, therefore, a matter of experience. This experience, and the resultant comprehension, is the result of the training and conditioning which occurs either external to the FPS game (for example, through exposure to popular commercial cinema sound conventions) or which takes place during initial exposure to the sonic conventions of FPS games in general or to the specific FPS game being played. These conditions apply equally to both the sound designer and the player who, ideally, should have similar socio-cultural experiences and understandings of sound. FPS game sounds may be described as

---

[1] That is, a particular use and disposition of sound as a defining feature of the subgenre.

a set of sonic signs or auditory icons which may be analyzed through semiotic terminology, such as indexicality, iconicism, symbolism or metaphor for example, in an attempt to explain how intended meaning is (ideally correctly) translated to received meaning (see chapter 4).

The FPS game engine may be understood as a sonification system in which sounds are (re-)encodings of non-audio data. This game world data may derive either directly from the game engine, as in the case of game status sounds for instance, or, in the majority of cases, is an expression of player activity, such as the sounds of footsteps or the firing of weapons.[2] In most modern FPS games, sounds are audificated from stored audio samples. There is also likely to be a use of audiation by players, especially experienced players, who may, on the basis of prior experience and expectation, surmise information about game world objects and events from non-visualized sounds (see chapter 4).

Sounds in the FPS game have a virtual indexicality[3] to the game world's objects and events. Synchresis is one method to explain the perceptual conjunction in the game environment of image and sound as they emanate from different physical locations in the user environment (see chapter 5). Sounds in the FPS game are, therefore, virtually causal[4] and this causality is, in the FPS game, more concerned with action or event than with image (see chapter 5).[5] For all actions on the part of a player which produce a sound, that sound is, additionally, causal to that real-world action and so maintain, potentially, meaning derived externally to the game world. The use of discrete audio samples as an enabler for the phenomenon of virtual causality, providing immediate feedback to player input, has positive implications for player immersion and participation in the game.

All sounds, or the use of some sounds, in the FPS game contribute in some way to player immersion in the acoustic ecology and it is this immersion within and the player's creative participation in the game's acoustic ecology which, in large part, affords the perception of immersion in the FPS game world (see chapter 7). Thus, the player is physically immersed in the real resonating space and, through kinaesthetic techniques and the ability to produce a range of sounds through haptic input, is drawn into the virtual resonating space which is then meta-synchretically mapped to the visual game world and activity that are represented either on- or off-screen (see chapter 6).

---

[2] Many game status sounds are themselves a result of cumulative player actions.

[3] Which becomes increasingly real in its indexicality with increasing exposure to the game.

[4] Although I would strongly argue for the affective usefulness of more causally mysterious sounds.

[5] In the real world, the physical vibration of (or by a part of) the source object is the action which causes sound. Recording and reproduction break this directly indexical causality (see chapter 5). Yet, in the FPS game, synchresis is one explanation for the virtual causality of sounds to objects and events portrayed on screen.

8.2 The taxonomy

An initial taxonomy of FPS game sounds is one that peruses the classification of audio samples as found on either the distribution medium or on the installation drive.[6] This form of taxonomy provides useful insights into the sound designer's classification system which itself may be extrapolated to the meaning they intend for particular sounds. Yet this is a form of FPS game sound taxonomy which does not appear in any of the literature reviewed in chapter 2 other than brief mention of character sounds, interactable sounds or environment sounds for example (see Stockburger, 2003; Folmann, n.d.; Friberg & Gärdenfors, 2004). None of the literature explicitly examines the distribution or installation media for clues as to the sound designer's intentions. (See Appendix 1 for a summary of *Urban Terror* audio samples classified by installation directories.)

At the very least, this taxonomy provides a division between diegetic and nondiegetic and there is typically a separate directory for music or menu interface audio samples as opposed to other audio samples which themselves may be sub-classified into character, interactable, environment or feedback audio samples. Thus, of the 607 base audio samples of *Urban Terror* (game-specific audio samples as opposed to level-specific audio samples), fully 601 are available to be used during gameplay[7] with the remaining six being the menu music (one) and menu interface sounds (five). The 601 audio samples are, therefore, diegetic whilst the remaining six are nondiegetic.

The game designer-constructed organization of audio samples in *Urban Terror* is illuminating in several respects. Firstly, it is an indication of how the game code deals with sound and its relationship to a variety of characters, objects and locations within the game. Sounds that players' characters create as they move, fire or taunt are separated from environment sounds which are part of a location; sounds of interactable objects are separated from the sounds non-interactable objects make and diegetic sounds are separated from nondiegetic sounds. Secondly, the sheer number of sounds is an indication of the importance of sound to the game experience. Thirdly, this organization of sound indicates some of the technical limitations of the game namely in the areas of media storage and computer memory. As an example, some audio samples of footsteps are shared between the characters and this decreases the number of sounds which must be stored on the game's distribution medium (a compact disc in this case) and which must be loaded in the computer's memory while playing.

As a taxonomy, though, it says little about the function and meaning of sound, how sound is used in the game by the player or how sound helps to shape the FPS game acoustic ecology. Nor is it possible to successfully combine it with the terminology of the conceptual language summarized above as this, in the main, deals with the player experience of and engagement with sound. Other taxonomies must be derived and combined with the conceptual language to achieve this.

---

[6] For those FPS games that are formatted for consoles, the former is the only option. For those to be installed on personal computer hard drives, the latter is the better option as the installation program may re-order some audio sample directories. Additionally, in this case, game updates may not always include the original audio samples from the base game but may only include further audio samples to be added to those already installed. The list of audio samples in *Urban Terror* given in Appendix 1 is taken from the installation directory.

[7] Which ones will be used or not depends, for example, upon the game mode and which characters or weapons are being used. Certainly, not all audio samples will be sounded during any one playing of a game.

Before proceeding to these other possible forms of classification, those that elucidate to a greater extent the player experience of and engagement with sound, a discussion of the means of sound creation and production at the game design stage proves useful. This is not least because taxonomies derived from this shed light on the degree of interaction possible in the FPS game which directly relates to the player immersion within and participation in the acoustic ecology. In any digital game, sounds heard during gameplay and from within the game environment may be synthesized or digitally recorded and stored as a discrete audio sample. In all modern FPS games, most such sound consists of audio samples and this is certainly the case for *Urban Terror* (see chapter 5 for a discussion of the limitations of such a practice and the potential benefits of the real-time synthesis of sound). Most of these audio samples are sounded in response to player input, game status (which, in most cases, is an indication of player activity[8]) or bot activity in games where bots are employed. A smaller number of environment audio samples are under the control of the game engine although their sounding may be responsive to player location.

The abundance of recorded audio samples (as opposed to synthesized audio samples), which may be described as nomic auditory icons (chapter 4), combined with their appropriate in-game use (in other words, they are causal sounds with a high degree of virtual indexicality[9]), is a good indication of the level of realism[10] the FPS game aspires to. *Urban Terror*, which is usually described as a realism mod, is a prime example; of the 601 diegetic audio samples available, the only synthetic audio samples (the only symbolic auditory icons) are those related to game status events, such as when a flag has been captured. This may be compared to *Quake III Arena* which, set as it is in a more fantastical gamescape, has a greater proportion of symbolic auditory icons representing not just game status events but also various audio samples sounded by player input (such as those to do with power-ups and teleporters).

Over 80 of these non-synthesized samples in *Urban Terror* may be initially classed as somewhere between a metaphorical or a caricature nomic auditory icon especially to the inexperienced player. These are the audio samples the meaning of which it takes time to learn, more so than the almost immediate inference to be gained from nomic auditory icons with their greater articulatory directness. An *Urban Terror* example is the audio sample representing healing. Apart from some indeterminate clicking sounds at the start of the sample, it sounds like a tight roll of cloth being quickly unwound so may be viewed as a synecdoche standing for the act of bandaging a character's wounds. With increasing experience and immersion within the game world, as the player becomes the game character in the game world, these audio samples become increasingly indexical to the actions and objects they represent and so become increasingly nomic and less metaphorical.

The 247 recorded speech audio samples in the *Urban Terror* sound/radio/ directory are not auditory icons as Gaver (1986, p.168) defines them because dimensions of the sound source[11] are not used to give information about the object or action; instead language is used. They therefore stand in a class by themselves maintaining

---

[8] An exception being any signal indicating the length of time remaining to play in timed games although, even here, the argument could be made that this is a sound initiated by the presence of players at the start of the game.

[9] For example, a recording of a shot-gun is sounded each time a shot-gun is fired.

[10] Realism as in simulation approaching emulation.

[11] That is, the material of the object or the springiness of the action, for example, are not sonified.

a strong indexicality with the original recorded voice yet, at the same time, coming to represent not the human behind the microphone but the human behind the character not least because they must be actively keyed by the player to be heard. Furthermore, each player has a limited set of audio samples which can be triggered from the computer keyboard[12] and so chooses his favourite radio messages thus helping to define his in-game persona further.[13] For any one player, his own radio messages played to team mates become, like the sounds of his breathing, part of his proprioceptive system — a vocal extension of themselves into the game and the only method with which to vocalize within the gameworld.[14]

All audio samples may be classified as either diegetic or nondiegetic following film sound theory (see chapter 7). However, differences in the creation and resultant nature of the FPS game soundscape compared to the film soundscape require refinements of the term diegetic. All sounds in the FPS game consist of discrete audio samples and, unlike film, there is no complete game soundtrack which is stored on the distribution medium and played during gameplay. As noted above, the FPS game soundscape which forms a part of the acoustic ecology is created in real-time through the agency of game actions (the sounding of game status feedback or environment audio samples for example) or through the agency of player input acting upon the discrete audio samples which form the soundscape's palette. Furthermore, with any playing of the game (even the same level), the resultant soundscape will be substantially different (see chapter 5 for a discussion on reproduction versus originality in the case of the soundscape) for the one player and, in a multiplayer game, the soundscape experienced by one player will also be substantially different to that simultaneously experienced by other players. It is for this latter reason that I define the terms ideodiegetic (those sounds that any one player hears) and telediegetic (those heard and responded to by a player — they are ideodiegetic for that player — but which have consequence for another player; they are telediegetic for the second player). Furthermore, ideodiegetic sounds may be classified as kinediegetic (sounds initiated directly by that player's actions) and exodiegetic (all other ideodiegetic sounds). Such a taxonomy, then, provides a foundation from which to explain player involvement in and experience of the FPS game acoustic ecology in a way in which film sound theory definitions cannot.

Of the class of diegetic audio samples, and in the context of a multiplayer game, all global feedback sounds (such as game status messages) may be classed as exodiegetic sounds. They are ideodiegetic in that they are heard by all players (simultaneously) but are initiated by the game engine in response to significant events.[15] Thus, in *Urban Terror*, the symbolic auditory icons used to indicate the capture of a flag or the sound of the bomb exploding in bomb mode are exodiegetic sounds. All other audio samples may be ideodiegetic or telediegetic depending upon context. These include environment sounds (which are usually level-specific audio samples rather than game-specific audio samples) such as the Bach organ fugue in the *Abbey* level. All audio samples which are, for example, weapons-related, are

---

[12] There are more radio messages supplied than there are available keys especially after some keys have been assigned to other functions such as weapon selection and movement. *Urban Terror* does enable further radio messages to be accessed through the in-game menu but this can be cumbersome to use.

[13] To a certain extent, the selection and frequency of use of such radio messages is a key to the persona of the player behind the in-game character — possibly a subject for future research.

[14] With the exception of Voice over Internet applications which are not part of the *Urban Terror* download.

[15] Although such events may arise as a result of combined player actions.

radio messages or are the sonic 'personality' of any particular character may, if they are ideodiegetic, be either kinediegetic or exodiegetic in that if they are sounded directly as a result of player action (the firing of the *SPAS* shotgun, for example), they are kinediegetic for that player but are exodiegetic for other players within hearing range. If such sounds have consequence for other players who do not hear them (for example, the blast of the shotgun which kills an enemy may draw others of his teammates to that location which itself may provide opportunities for the opposing team), they may also be classed as telediegetic for these other players.[16]

As has been noted by several writers cited in chapter 2 (Stockburger, 2003; Friberg & Gärdenfors, 2004, for example), sound in the FPS game may be attended to in one of three modes: reduced listening; semantic listening and causal listening. Reduced listening, as noted by Stockburger (2003), is little used by experienced players (see the discussion below on perceptual retainers). What none of these writers suggest is that the mode of listening may change depending upon context and experience (see chapter 4). Furthermore, a fourth mode, navigational listening, has been identified. This is required because of the unique (compared to electro-acoustic music and film sound theory where the original three modes were first described) abilities of the FPS player to move his character around the 3-dimensional game world.[17]

Keynote sounds are those audio samples which form part of the sonic ambience and which may not be directly triggered by the player being, instead, sounded by the game engine.[18] An example in *Urban Terror* is the Bach organ fugue in the *Abbey* level or, in the same level but in a different location, the twittering of birds. There is some ambiguity here which is not captured by the monolithic descriptions of such environment or ambient sounds in the literature discussed in chapter 2. The player does typically have some kinaesthetic control over the sounding of these sounds; by simply moving away from that location, the sound may be attenuated to silence (and vice-versa). Furthermore, if a keynote sound is a sound which is not intended to be consciously listened to, merely forming the background for more perceptually important sounds, the decision to consciously attend to a sound or not is often a matter of player choice. As a trained musician, I have always found it difficult to ignore elevator music or muzak (or indeed any music I hear) and always set about listening to it, analyzing it, critiquing it or kinaesthetically responding to it; so it is with the organ fugue in *Abbey*. Additionally, though, every time I hear it, it brings to mind the baptism/execution montage towards the end of *The Godfather* (Coppola, 1972). In this sequence, the Godfather is attending the baptism of his grandaughter, swearing to be faithful to God and the laws of the church, in a scene which is intercut with scenes of his mafiosi gunning down rival gang bosses. This montage is held together by the diegetic organ music from the baptismal cathedral.[19] Thus, I cannot help consciously attending to this music and, while a keynote sound may still be consciously attended to (as it might be at the first playing of a game or level before it

---

[16] Thus, the duplet of ideodiegetic and telediegetic properties of an audio sample may be tenuously likened to the Heisenberg uncertainty principle (the Heisenberg indeterminacy principle) in which a pair of observable conjugate quantities (ideodiegetic and telediegetic) of a single elementary particle (the sound) cannot be measured with precision. An ideodiegetic-telediegetic duality akin to the wave-particle duality of physical phenomena. Perhaps this is an instance of the Observer Effect or, in this case, the Listener Effect.

[17] For instance, the previously used example of the Bach organ fugue in the *Abbey* level of *Urban Terror* as audio beacon or the use of navigational listening to track the source of the sounds of distant battle.

[18] As explained in chapter 6, they may be triggered by other players but are judged by the one player to be distant and of little interest and so form part of the general ambience of battle.

[19] Although, as the music is played over the killing scenes it may be described as nondiegetic.

subsides into the background), this Bach fugue, because it only plays in one particular game locale in the *Abbey* level, always draws me to that locale.[20]

A signal sound is a foregrounded sound which is designed to be consciously attended to because it potentially contains important information encoded within it. Most of the game-specific audio samples in *Urban Terror* may be classified as signal sounds when they are sounded in a context which foregrounds them. Thus, the loud, and therefore proximate, sounding of gunshot samples are worthwhile paying attention to (particularly in the individual deathmatch game mode). However, if the sounds of battle are distant, they may be classed as keynote sounds particularly if they are relatively constant and the player's attention is directed elsewhere. All game status indicators and team radio messages are signal sounds because, although they are as pervasive as keynote sounds, they are usually louder and therefore more proximate and, in the case of radio messages, have no reverberation, thereby foregrounding them through the lack of depth cues.

Soundmarks are identifying aural features of the *Urban Terror* environment and may be either signal sounds or keynote sounds which are consciously attended to. For a particular level, these are level-specific sounds, as is the case with the Bach organ fugue which does not appear in any other level and so is an identifying aural feature of *Abbey* and they can, as in this case, function as a soundmark, an audio beacon for navigational listening. They can also be game-specific sounds, particularly team messages and game or flag status indicators; any sound which is unique enough to the game for a listener, upon hearing the sound (even if it has been taken out of context), to identify it as part of *Urban Terror*. Symbolic auditory icons, such as flag status signals, are more likely to be uniquely identifying of an FPS game than nomic auditory icons because the latter are derived from recordings of existing real-world, and therefore external, sounds.

FPS game sounds may be categorized according to a variety of immersive principles as outlined in chapter 7. Following Ermi and Mäyrä's (2005) ideas, all FPS sounds can contribute to sensory immersion where the sounds of the game world override those in the player environment. It should be noted, though, that the level of sensory immersion is dependent upon a range of factors beyond the control of the game designers including the relative loudness of the two sets of sound and the audio hardware used — one of the factors influencing the decision of most FPS players to use headphones (Morris, 2002) is likely to be a greater sensory immersion. Many sounds offer challenge-based immersion by requiring a response which includes the use of both mental and kinetic skills. These, in *Urban Terror*, include a range of weapons fire necessitating a threat assessment on the part of the potential target (which may be based upon the localization and depth affordances offered by the sounds and the ability to precisely contextualize oneself in relation to team members and the enemy, for example) and, if the sound poses a threat, necessitating decision-making leading to kinetic action (to attack or to escape, for example). It is typically the case that these sounds are ones which are produced by other players and they usually relate to actions involving weapons. However, whilst most level-specific environment sounds in *Urban Terror* generally do not offer challenge-based immersion possibilities, audio beacons (such as the potential first use of the Bach organ fugue in *Abbey*) require the navigational listening mode and, therefore, mental

---

[20] To the *Abbey* neophyte, the music may be used as an audio beacon (in the navigational mode of listening), very deliberately focussing on that sound as a means to learning the layout of the level.

skills.

Sounds offering imaginative immersion possibilities are those which help the player identify with his character and the game environment and action. In the first case, *Urban Terror* offers a range of character sounds some of which (such as the character's breathing whose rate varies according to the exertions of the character) may be classed as proprioceptive sounds and which, with a high level of immersion, may be seen as aural prostheses similar to the prosthetic limbs seen receding into the screen. Exteroceptive sounds affording imaginative immersion through identification include a range of sounds which aid in contextualizing the player's character within the environment — in *Urban Terror*, an example would be the thud of the body landing on the ground which is dependent upon the effect of the simulated gravity and produces a louder sound than that of footsteps (the fall is also likely to lead to a reduced health status).

McMahan (2003) categorizes digital game elements as perceptual sureties, surprises or shocks. The latter militate against immersion in the game world by being external stimuli (or errors in the game) that remind the player that this is just a game taking place within the player's real-world environment. Sureties are mundane cues — expected details providing an experience which is consistent with the rules and conventions of the game world. In *Urban Terror*, the creaking of a door as it opens or closes or the footsteps of a player moving around are aural examples of this. Surprises, according to McMahan consist of three types; attractors (inviting the player to do something), connectors (helping player orientation) and retainers (causing the player to linger in game world locations) (pp.75—76).

A variety of sounds in *Urban Terror* fulfill these requirements. Indeed, any sound inviting an active response may be said to be an attractor. Thus, the sound of gunfire in the distance may tempt the player to investigate and team radio messages detailing enemy actions invite a response on the part of the team player. Many sounds, particularly environment sounds, function as connectors and they are often attended to in the navigational listening mode (the Bach organ fugue in *Abbey* being a good example). Locational and depth properties are important parameters of sounds functioning as connectors. Although, at first playing of the game, the player may derive enjoyment out of certain sounds and so may linger in a particular location in order to hear more, the nature of the FPS game is such that more-or-less continual movement is required of the player to seek out or avoid the enemy or to attack the enemy base and so, for the experienced player, no sounds in *Urban Terror* may be said to be retainers.[21]

The perception of a variety of spaces is one of the main contributing factors of FPS sounds to the perception of, and immersion within, the game world. In terms of resonating spaces, there is a real resonating space, which is the acoustic volume enveloping and morphing around[22] the player, and a virtual resonating space, matching the illusory visual space depicted on the screen,[23] the perception of which is created by parameters of sound such as localization, depth cues and

---

21 For those who enjoy sound, enjoyment of the complete collection of sounds and the use of that sound in any particular game may be part of the reason to continue playing the game in which case they may all be classed as retainers.

22 Similar to the semiotic space morphing around the character in Innocent's (2003) game *Semiomorph*.

23 Other virtual spaces may be identified as separate volumes within the game world.

reverberation. Such cues may be processed in real-time with more sophisticated game audio engines or they may be encoded into the audio samples on the distribution medium (as happens with *Urban Terror*). Sounds providing this affordance are choraplasts. Sounds may also function as topoplasts where they create the perception of paraspaces such as locations in the game — again, the Bach organ fugue in *Abbey* functions in this way by working with the visual depiction on screen to provide the location of a Gothic church and its environs (primarily, a religious paraplace). Additionally, sound may provide the affordance of the perception of time passing or of a particular temporal period in the past, present or future. The former are chronoplasts and, because sound is vectored through time, that is, it takes time to hear a sound, all sounds have a chronoplastic function. The latter are aionoplasts and, in *Urban Terror*, the weapon sounds have this function setting the game in the modern era (rather than, for example, the mediæval age in which the sounds of automatic rifle fire would be unknown). However, specific sounds, usually environment sounds attached to particular levels, may provide the perception of more specific periods. The crackling Edith Piaf song playing on the 1940s gramophone in the ruined French village of the World War II level *Rommel*[24] is an apposite example.

8.3 Conclusion

This chapter has consolidated and summarized the conceptual framework formulated throughout the previous chapters for the purposes of explicating the hypothesis of the FPS game acoustic ecology in chapter 9. Although it has primarily used *Urban Terror* for illustrative purposes, the conceptual framework may also be utilized for the analysis of player and sound relationship to be found in other FPS games. I suggested in section 8.2 that the ratio of nomic auditory icons to symbolic auditory icons may be used to differentiate realism FPS games from the more fantastical FPS game type. Indeed, components of the framework may also be used for sonic analyses in other digital game genres. For example, it may be that the concept of telediegesis can be used to help understand aspects of player relationships in MMORPGs.

The model of the acoustic ecology, slowly built up throughout this work, may now be presented in the complete diagram of *Figure 8.1*.

---

24 Modeled on the village of Ramelle depicted in the climactic scenes of *Saving Private Ryan* (Spielberg, 1998).

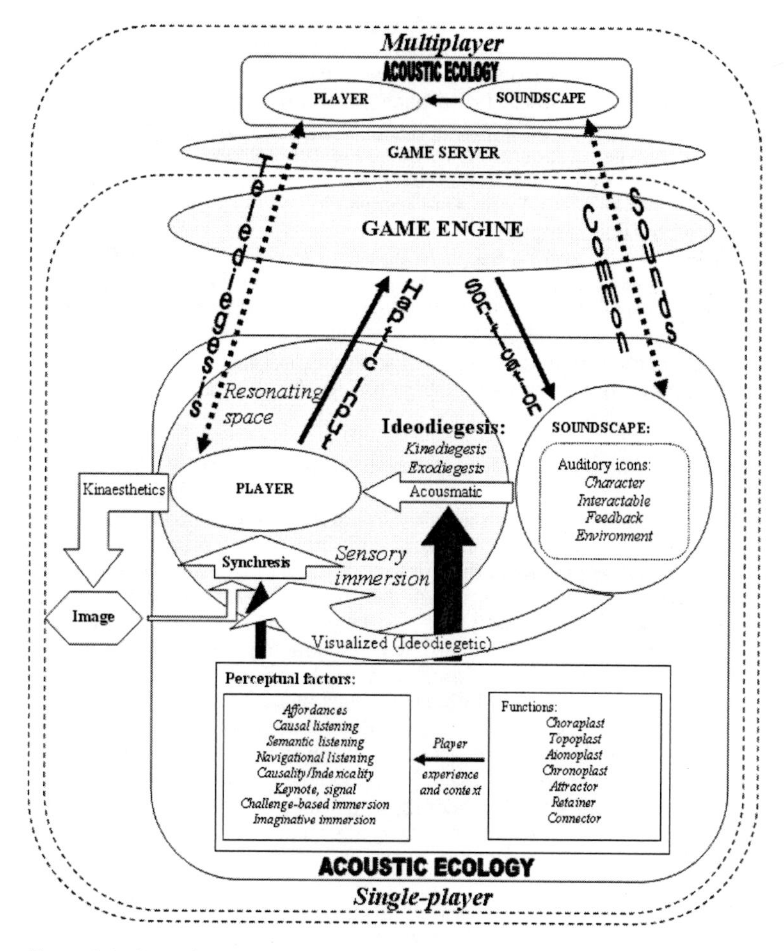

Figure 8.1. A concise conceptual framework of the FPS game acoustic ecology in multiplayer mode.

This model exhibits all the elements of the conceptual framework described thus far. Because it is a model of an *acoustic ecology*, it importantly shows relationships between player (the listener) and soundscape and thus fulfills the requirements of an acoustic ecology as explained in chapter 1. As the acoustic ecology of the FPS run and gun game, though, it includes a variety of components and relationships which are unique to digital games[25] such as the game engine, image, a range of spatial and immersive elements and perceptual factors and, because this is a model of a multiplayer game acoustic ecology, the game server and other players and their

25 Some of which may be unique to the subgenre. This is a matter for future research.

soundscapes.[26]

There has been little research into the player relationship to digital game sound and the literature review of this work not only demonstrated this but identified numerous gaps in existing theories which fail to account for one aspect of this relationship or another. Consequently, chapters 3 through 7 utilized a multidisciplinary approach to fill these gaps arriving at a more inclusive and comprehensive conceptual framework for the study of FPS game sound than has previously been attempted and, ultimately, developing the model of the FPS game acoustic ecology as presented in *Figure 8.1*. Although the conceptual framework and the model are focussed on FPS games as their paradigm, it may well be the case that they (or aspects of them) may also be used to analyze the wider area of digital game sound in the future. Consequently, although the following is a digest of the framework and model's contribution to the study of FPS game sound and the player relationship to that sound, it may also be viewed as a contribution to the study of digital game sound in general with the caveat that much further research and testing (using different digital game genres) is required.

The conceptual framework and model both include and modify some aspects of existing digital game sound theory and expand upon it in a number of significant ways. It is important to take into account the game sound designer by investigating not only methods of sound production and design but also by examining the organization of audio samples in the game engine and this insight led to a taxonomy of audio samples as shown in the model of *Figure 8.1*. This approach is strengthened by the identification of these audio samples as auditory icons (because auditory icon design is intimately connected to the design of meaning in sound) and to the noting of the role of sonification in the creation of both the soundscape of the acoustic ecology and the sonic, relational framework within which the player operates. These insights were founded upon a fuller investigation of the potential meaning of sound in the FPS game and led to the identification of a range of perceptual factors modifying the comprehension of sound within the game context. In particular, I have been able to adapt existing theories of immersion in virtual environments, to identify the role of player experience and context in assessing the meaning of sound and to contribute new terminology (such as *navigational listening*) to current concepts.

As created and contributed to by sound, a range of spaces has been uncovered in the FPS game. The new concept of *resonating space* has implications for player immersion in the acoustic ecology and, therefore, the game world. The abilities of sound to create resonating spaces and to contribute to the perception of paraspaces are accounted for by the defining of new functions of sound: *choraplastic*, *topoplastic*, *aionoplastic* and *chronoplastic*. The existing concept of synchresis between image and sound has been adapted to account for a meta-synchresis between image and resonating spaces in the FPS game. This too proves useful to an understanding of player immersion. Furthermore, it has been argued that sound in the FPS game is, in the main, predicated upon player action and this strengthens the argument for player immersion within and participation in the acoustic ecology. In addition to the use of synchresis and the causality of sound, the term *first-person auditor* has been defined in order to account for the central position of the player in the FPS game world and some of the perceived relationships between image and

---

26 For clarity, only one other player and soundscape is shown here.

sound.

The concept of diegetic sound has been borrowed from Film Theory and developed in order to explain, firstly, player participation in the acoustic ecology through the triggering of audio samples and, secondly, the sonic relationships between players in a multiplayer FPS game. To this end I have proposed new terminology *viz.* *ideodiegetic sound* (comprising *kinediegetic sound* and *exodiegetic sound*) and *telediegetic sound*. Lastly, a diagrammatic model of the acoustic ecology of the FPS game has been proposed, containing the main points of the conceptual framework, that will be used in chapter 9 and that may continue to be used for future research into the nature of the player and sound relationship in FPS games. Thus, the conceptual framework and model greatly develop and expand upon existing theories of digital game sound as discussed in chapter 2.

The following chapter uses the conceptual framework and the FPS game acoustic ecology model of *Figure 8.1* in order to provide a full statement of the hypothesis of the FPS game acoustic ecology. In doing so, I also explore the use of autopoiesis as a way to account for a multiplicity of acoustic ecologies within the multiplayer FPS game and this process leads to further insights on player immersion. To complete the chapter, aspects of the model are utilized to explain the FPS acoustic ecology as dramatic performance. This is only one of a variety of conceptualizations of the acoustic ecology which may be attempted and has been chosen in order to demonstrate the robustness and versatility of the model.

Chapter 9

The Acoustic Ecology of the FPS Game

Introduction

This chapter provides the full exposition of the hypothesis that the sound of the FPS game may be described and analyzed as an acoustic ecology in which the player is an integral and contributing component. The hypothesis depends for its explication upon the conceptual framework that has been developed in the preceding chapters and summarized in chapter 8. This framework forms the foundation of this chapter which explains the acoustic ecology of the FPS game in the general case while simultaneously using *Urban Terror* (Silicon Ice, 2005) as an example of the acoustic ecology in the specific case.[1]

The first section presents a description of the main components of the FPS game acoustic ecology and discuss the relationships between them, elucidated by the conceptual framework and illustrated by examples from *Urban Terror*. This is a description of one player's acoustic ecology and from that player's point of view. However, the model presented at the end of chapter 8 indicates the existence of other acoustic ecologies in a networked, multiplayer FPS game and so the second section presents the case both for multiple player acoustic ecologies and for a virtual acoustic ecology of the multiplayer FPS game and conceptualizes this through autopoietic theory. The final section provides a test of the conceptual framework and model by utilizing them to explain the Prelude of chapter 1 as a dramatic performance. This is an attempt to demonstrate the robustness and versatility of the acoustic ecology model for the analysis of FPS game sound from just one perspective.[2] As in the previous chapter, the multiplayer FPS game is the focus.

9.1 The main components of the acoustic ecology

This section comprises a description of the FPS game acoustic ecology by identifying its major components, the relationships between them and the technical genesis of the ecology. It makes use of the conceptual framework of chapter 8 and illustrates the description with examples from *Urban Terror*. In the case of the FPS game, the basic building blocks of the acoustic ecology are players and the soundscape. It is important to note that these components will vary even for repeat playings of the same level in the same FPS game. Players continue to deepen their experience of the game and its sounds and one player will have different behaviour patterns or schemas to another favouring, for example, a different set of radio messages or weapons. Furthermore, some audio samples which are available to be triggered in the level by players or the game engine may not all be used. This is more likely to be

[1] The images of soundscapes from *Urban Terror* which are presented here are taken from multiplayer sessions held over several weeks in which the group of players was a mix of those with much experience in FPS games and those with little experience. Only two had played *Urban Terror* before.

[2] There are many other perspectives with which the model may be tested in the future, theories of social networks or chaos theory, for example, but space does not permit a wider survey.

the case for game-specific audio samples than level-specific audio samples, and is especially the case for team radio messages, although it may also be the case for audio samples attached to particular weapons which may no longer be in use. Thus, it is not just that the acoustic ecology changes during the gameplay, as I describe below, but it is also the case that the components of that ecology, the players and the soundscape (through its available palette), may be different from the outset at each playing of the game.

Upon spawning in a multiplayer, capture the flag configuration of the *Urban Terror* level *Abbey* (see the Prelude of chapter 1), I experience sound which I have either triggered myself or which derives from other sources such as other players or the game environment. As an experienced player on this particular level, I am aware of which sounds have important affordances and which can be safely pushed to the periphery of my perception and cognition. This is not to say that the latter sounds have no significance. Indeed, such keynote sounds as the pervasive birdsong, the localized organ music or the sounds of team members readying themselves for action provide important immersion cues, localization cues and game event cues. These all aid in creating the stage on which I can signal my presence to other players through the triggering of kinediegetic sounds.

The first few seconds of any capture the flag game in *Urban Terror* are sonically similar, comprising environment sounds (if any) and the sounds of teammates, because teams spawn at separate points in the level and must move towards the opposing teams' bases for significant action to occur. Thus, as I move towards the enemy's base intent on stealing their flag, I begin to hear sounds which have potentially more significant gameplay affordances for me. For example, the exodiegetic sounds of battle in various directions indicating that some members of my team have already encountered the enemy. Such sounds in the soundscape may provide the affordance of acting as audio beacons guiding me through the navigational listening mode, if I so choose, to those points of action or I may use the sounds for evasive action instead. Importantly, though, they are signal sounds indicating player presence and activity; other components of the game's acoustic ecology contributing to my soundscape. No less important are my own kinediegetic sounds. Prior to meeting the enemy in *Urban Terror*, such sounds will be the sounds of my footsteps and breathing. The latter is particularly important; its rate and intensity increases as I expend energy in running through the level and these increases are matched by a decrease in my speed of movement. Experience informs me that I should conserve energy in order to be able to run faster when the exigencies of combat require it.

The game engine contributes to the soundscape not only with the keynote sounds of birdsong and organ music but also with signal sounds to which I must be attentive as they are indications of potentially significant gameplay activity. In an *Urban Terror* capture the flag game, the alarm indicating the taking of a flag is a key signal sound and, from experience, context and various on-screen displays, I can recognize which flag has been taken and react accordingly. In this case, I know that it is one of my team members who has snatched the enemy flag and my reaction is to head towards the enemy base in the hope of meeting and aiding that player. As I do so, enemy combatants have been following the flag-carrier through the twists and turns of the level by following the sounds of his footsteps and the explosions of the grenades he lobs behind him as he flees. For me these are telediegetic sounds at first (until I am in close enough proximity to hear them as exodiegetic sounds), and their consequences are such that, when I finally meet up with my teammate, I discover a posse in hot pursuit. It is at this point that I am able to signal my presence to the enemy not merely visually (for those who can see my character on screen) but also sonically and acousmatically through the sound of my footsteps and, more significantly, through the sounds of weapons fire which I am able to trigger at will. Likewise, I am aware of their presence and activity in the game world through both their visualized and acousmatic sounds.

As a player, then, I use sound to orientate, contextualize and immerse myself in the game world. I also use it to understand the events of that world and to relate to other players. Yet this game world would not exist and be perceived by me without the presence and participation of at least myself. Similarly, the acoustic ecology would not exist without at least my contribution in addition to the sounds the game engine itself sounds. These latter sounds usually result from player actions, such as the flag alarm, and even the environment sound of birdsong requires me to perform a series of actions leading to my in-game spawning. Exposure to real-world acoustic ecologies and the sonic conventions of media such as film and other digital games are an aid to understanding the meaning of sounds in the game. Furthermore, through repeat playings of the game, I have gained experience in interpreting the peculiarities and specific meanings of the *Urban Terror* soundscape (that distinguish it from other acoustic ecologies) and am able to use this to my advantage during gameplay. Without me or other players, the soundscape does not exist. Without the soundscape, the game experience would be lacking and, as I am used to operating in a sonic environment, would be less likely to engage or immerse me within the game world or to be beneficial to my playing ability. Player and soundscape are symbiotic components of the FPS game acoustic ecology and are therefore analyzed here using the key concepts of the conceptual framework.

9.1.1 The soundscape of the acoustic ecology

Previously, the combined sounds generated during FPS gameplay have been referred to as the game's soundscape and it has been shown that soundscape terminology, as defined by Schafer (1994), may be applied to individual sounds within the game (see chapter 7). However, I have not yet analyzed the FPS soundscape in its entirety, limiting myself to a discussion of its component parts only. As the FPS game soundscape is analogous to the acoustic environment (that is, the sonic surroundings in which the players and bots are active), an analysis of that environment as it forms part of the acoustic ecology of the game is pertinent.

As it happens, there are two uses of the term soundscape. It was originally coined by Schafer and it is clear from his writings that a soundscape for him forms part of an acoustic ecology. As an example, "[t]he church bell is a centripetal sound; it attracts and unifies the community in a social sense" (Schafer, 1994, p.54) and he suggests that humans are shaped by their acoustic environment (Schafer, 1994, pp.9—10). These are indications that man-made sounds (as well as natural sounds) affect humans in addition to humans creating and changing the soundscape themselves. Such descriptions parallel Böhme's definition of an ecology as comprising a relationship between the environment and people. Among the terminology which Schafer uses to describe soundscapes in this sense are keynote sound and signal sound which have already been freely applied to the discussion of FPS game sounds in chapter 7 and have been summarized in chapter 8.

The second usage of the term soundscape is one which arises from the field of electro-acoustic composition and that is the soundscape isolated and abstracted (and sometimes composed) from its source environment to be played, for example, in a concert hall or as a backdrop to a tourist tableau, but always recorded and therefore fixed on the recording medium. In this state, the soundscape may no longer be affected by its original inhabitants[3] nor may it affect them in turn. It is not, therefore, an acoustic ecology yet it retains a usefulness in analytical terms. This section, then, discusses the soundscape as a recorded artefact before moving on to discuss it *in situ*, as one of the two main components of the FPS game acoustic ecology,

The FPS soundscape as acoustic environment has some aspects which are fixed and predictable and other aspects which are more volatile and less predictable. For any FPS level, the fixed aspects, to name a few, may refer to any real-time processing the game engine is capable of (thus, in a certain location of the game level, there will always be a predictable quality to the reverberation applied to the audio samples), may refer to level-specific sounds and their ambit (for example, the birdsong and the more spatially circumscribed Bach organ fugue in the *Urban Terror* level *Abbey*, a centripetal sound if used as an audio beacon in the navigational listening mode) or may refer to the palette of game-specific audio samples which is available. Therefore, in a capture the flag configuration of *Abbey*, it is possible to predict with decreasing, but still substantial, accuracy that a player will hear the birdsong and the organ music, that any game-specific audio samples heard will be unchanging in their reverberant quality, that a variety of footsteps, radio messages, weapons fire and explosions will be heard and that flag status audio samples will be sounded at irregular intervals, the frequency of which is dependent upon the capabilities of the teams.

The more volatile and less predictable aspects of the acoustic ecology relate to the timing of sound events, which is evidenced by the frequency and density of game sounds (the game-specific and game status sounds in particular). Furthermore, in a multiplayer game, each player's sound experience reflects their first-person auditor status and activity within the game world. That is, each player's sound experience will differ from the next depending upon where they are in the level, their level of activity, their team membership, choice of weapon and character, for instance. That sound experience also results from the sounds other players make and, in the tightly connected diegetic world of the FPS game, may also be a consequence of the activities of those players. In terms of the conceptual framework outlined above, it may be said that, in a multiplayer game, the soundscape heard by the one player comprises both kinediegetic sounds and exodiegetic sounds and that any of these ideodiegetic sounds may be sounded as a response to, and, therefore, as a consequence of, telediegetic sounds.

In terms of the density and frequency of occurrence of sound events, there may be a degree of predictability to this too, especially at the start of a game and at each player's in-game (re)spawning. In a capture the flag configuration in *Urban Terror*, the two teams usually spawn at the beginning of the game at opposite ends of the level away from their team's flag (and each player respawns similarly after their in-

---

[3] If there ever were any.

game death). There is a short period, then, of the sound of footsteps with little or no gunfire as each team member moves towards the expected focal points around the flags possibly accompanied by the sounds of weapons being cycled as the players prepare themselves for action. The density and frequency of gunfire (amongst other ideodiegetic sounds) will then increase until the player's death, at which point the cycle begins again. There are other predictable features which I analyze further below.

These elements of volatility and predictability are represented by the graphical soundscapes of FPS gameplay shown in the following diagrams.

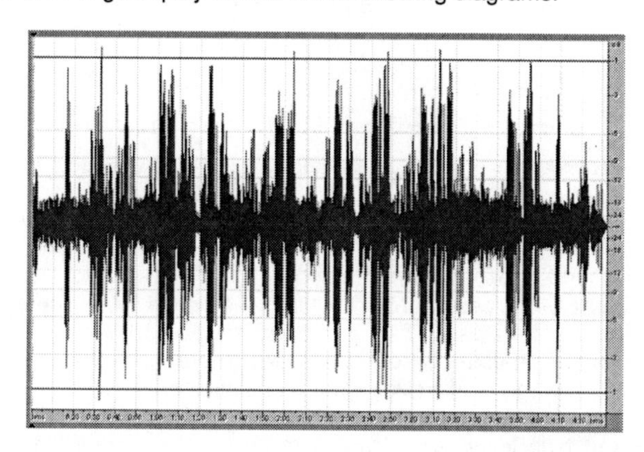

Figure 9.1. The soundscape of a free for all *Urban Terror* game.

Figure 9.1 is the soundscape of one of eight players in a free for all game in the *Urban Terror* level *Abbey*. The higher sound intensities shown are representative of high audio activity with the highest audio peaks indicating the player's firing and other kinediegetic sounds.[4] High levels of audio activity equate to high levels of activity on the part of the player or in the vicinity of the player as these lead to a high density of kinediegetic sound from that player or from other players in the area. Given the nature of *Urban Terror* and the type of game configuration, it is likely that these points represent dense bursts of weapons fire.

The next thing to notice about the diagram in *Figure 9.1* is that the soundscape not only graphically represents bouts of high activity but also indicates that these bouts are interspersed with periods of lesser or more distant activity. Assuming the player actively seeks out and engages the enemy, whatever the game configuration, this alternating model is predictable and may be applied to any FPS run and gun game. *Figure 9.2* is evidence of this.

---

[4] As with most FPS games, self-sounded weapons fire is usually the loudest form of sound event experienced by the player.

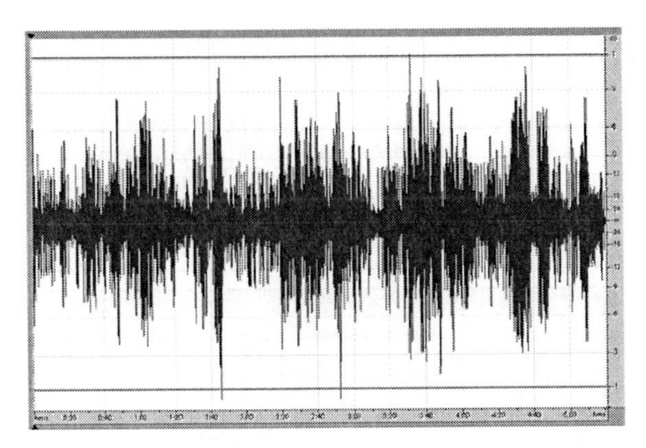

Figure 9.2. The soundscape of a capture the flag *Quake III Arena* (id Software, 1999) game.

Figure 9.2 is a graphical soundscape of a capture the flag game of *Quake III Arena* on the level *CTF3* with eight players as heard by one player. Like the free for all *Urban Terror* soundscape above in *Figure 9.1*, its peaks and troughs of audio intensity are indicative of peaks and troughs in the action within the game and that the player is likely to be aware of — soundscapes recorded from other players will show differences and, in some cases, similarities where the paths and actions of two or more players converge leading to common or similar sounds.

The frequency of high levels of activity is more random and less frequent for the capture the flag configuration compared to the free for all configuration. This is an artefact of the game configuration. As previously mentioned, capture the flag teams typically spawn at opposite ends of the level. Players tend to converge on the action hot spots (usually the flag bases or some point in the middle where teams' paths may cross) and so there is usually a period of calm as team members move towards the flags. This is demonstrated by *Figure 9.3* which is a soundscape of one player showing the first 14 seconds of a capture the flag game with eight players in *Urban Terror*.

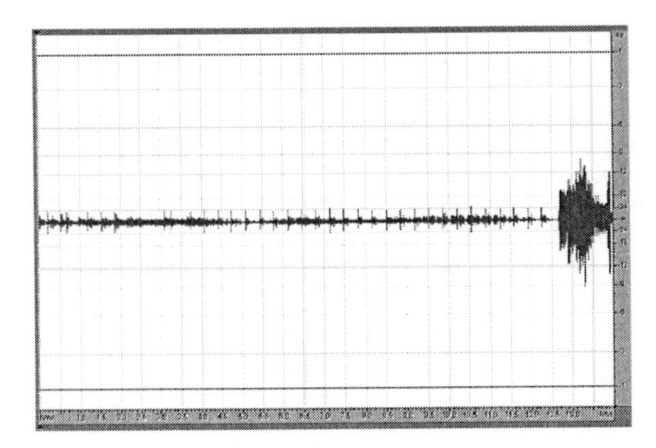

Figure 9.3. The soundscape of the first 14 seconds of a capture the flag *Urban Terror* game.

It takes about 13 seconds for the first gunfire to be heard by this player during which time the only sounds are environment sounds, the sounds of his own character's footsteps and grunts, his team members' characters' footsteps and grunts or equivalent vocalizations and the sounds of various items being picked up as players arm themselves.

This differs from the initial period for the free for all game shown in *Figure 9.1*. That being a free for all game there is, depending on the layout of the level, a high chance that the player will spawn next to, or in close proximity to, an enemy character in which case not only does the fighting start almost immediately but the player has less need to seek out the enemy. This perhaps, is one of the more illuminating uses to which these graphical soundscapes can be put. The less frequent the bursts of high audio intensity and the longer some of the quieter periods are, the more likely it is that the game configuration is one which promotes action hot spots which the experienced player, after spawning, will naturally gravitate to and will take time in reaching. Free for all games tend to have fewer or no such hot spots as the position of the player spawn points is more dispersed over the whole level. Capture the flag games tend to have hot spots of activity around the team flags which opposing teams must travel to. This is true for *Urban Terror* and *Quake III Arena* and may be predicted for other similar FPS games.[5]

Such combinations of volatility and predictability are common to the environments of all acoustic ecologies not just to those found in FPS games. As an example, sitting in front of this computer and at this time of day I have a very good idea of which sounds I will definitely hear and which sounds I am likely to hear. My computer fan will always be heard as will the rhythmic tap tap tap of my fingers on the keyboard. I know that in about 20 minutes or so, I will hear a train's whistle in the distance followed by the approaching then receding rumble as it crosses the nearby railway bridge. Birdsong (although of what type and what frequency is less easy to predict) will be heard. Less predictable sound events may be the sound of a police siren on the nearby road or of someone knocking on my door.

Returning to the FPS game, these individual sound experiences are still the sum of interactions between the game engine and the players that are part of the gameplay and so such experiences are a manifestation of the relationships forming the game's acoustic ecology. There are as many individual sound experiences as there are players in the game and each sound experience, being a part of the game's larger acoustic ecology, overlaps with others, potentially and telediegetically influencing other players and so affecting their ideodiegetic sound experience. Because each player's sound experience behaves like an acoustic ecology, they may themselves be termed acoustic ecologies (I discuss this further below).

The sound events (that is, each single sounding of a sound in the game) which make up the soundscape of each player's individual acoustic ecology result from level-specific environment sounds, game-specific status messages and other sounds triggered by that player or by other players. Sound events which are common across some soundscapes in the game will nevertheless sound differently for each player

---

[5] In terms of designing a game, game designers may use this insight to help analyze gameflow and activity in game level prototypes through the analysis of game soundscapes which may be seen as representative of the joint or divergent contemporaneous action taking place in multiplayer FPS games. Presumably, the bouts of intense activity should be well-enough interspersed to allow for the ebb and flow of adrenalin but not too dispersed to lead to boredom.

(with the exception of global sounds such as game status messages and global ambience); typically, this is a matter of different intensities of the sound for each player but the sound may also be transformed or masked by conjunction with other sounds. Furthermore, these common sounds, for example a player's footsteps or gunfire, which exist in the soundscape of other players and which arise from one player's actions, may have an ideodiegetic or telediegetic affect upon these other players' actions thus morphing the latter players' soundscapes and thereby providing new affordances to the players (potentially affecting their actions in turn).

Any FPS acoustic ecology must make use of choraplasts and topoplasts and, to a lesser extent, aionoplasts as the foundations of the spatial and temporal elements of the acoustic environment that is a component of that ecology. In many cases, these are the keynote sounds which, as Schafer describes, are ubiquitious and pervasive background sounds and, therefore, are environment sounds. They may well be consciously listened to by the neophyte or the analyst but they are not intended to be and usually they are not once the action gets under way and there are other sounds to attend to. These keynote sounds set the scene of each location in the game by the provision of cues indicating resonating spaces and paraspaces such as volumetric space, location and temporal period and are not kinediegetic sounds.[6] This helps their subsidence into the background; the fact that they may not be triggered by any player means that they provide few affordances as to the game action and so may be safely ignored. The majority of sounds which warrant attention in FPS games (as is the case in *Urban Terror*) sound as a result of player actions.

Subconsciously, though, such keynote sounds are the base of the acoustic ecology's soundscape giving it shape and depth and providing a more or less static matrix upon which signal sounds may be pricked out in patterned affordances. This Gestalt ground usually works in tandem with the image on screen (thereby forming a part of the entire game ecology[7]). It is, in the main, the sound which gives the image body but the image which confirms the material form of the sound yet both also provide some elements of resonating spaces and paraspaces. Thus, the lonely, reverberant call of a bird heard over a softly moaning wind without being in the context of an image may exist within a ghost town, a marine littoral or anywhere that is outdoors where that bird may be found, and which has large, reflective spaces. The image on screen provides the precise location to this audiation which may be in agreement with one of a number of possible scenarios which the player has visualized or may be something entirely different. If, as Bussemakers and de Haan (1998) suggest, sound is processed more quickly than image, then the image on screen cross-modally confirms what the ears have already processed.[8]

Against this sonic ground, aural figures (signal sounds) move inviting the player to consciously attend to and interpret their affordances. Typically, these signal sounds bear information which is germane to the game action (rather than to the game's environment). As an example, they may be speech, and this is borne out by the typical use of speech in FPS games; where it is intelligible, it is something to be

---

[6] Although, in some cases, they may be kinaesthetically controlled by moving toward or away from that locale.

[7] In the same way as the game's soundscape, its players and the relationships between them may be seen as an acoustic ecology, the game's actions or, extending this further, the whole game may be seen as an ecology. This could usefully be the subject of future research.

[8] Pudovkin (1934) notes this too when arguing for an asynchronicity of sound and image in early cinema.

consciously attended to whether it takes the form of in-game and game-provided instructions, hints to the player or game status messages or whether it takes the form of radio messages between team members as in the radio messages of *Urban Terror*. Signal sounds may also be global feedback sounds which are available to all players (the game or flag status messages in *Urban Terror*). Sounds which may be classed as symbolic auditory icons are also typically used as signal sounds as evidenced by a wide range of sounds, particularly power-up indicators, in *Quake III Arena* for example.[9] Signal sounds, though, may also be sounds, other than speech, which have a high level of virtual indexicality with their associated in-game objects and actions. Most of these virtually causal sounds are triggered by players — kinediegetic character sounds such as footsteps and gunfire — and, being interactable and biotic, betray the presence of a character to other players listening. It therefore pays to grant these sounds the status of signal sounds although, with sufficient distance from the action and source of the signal sound, they may in fact tend towards keynote sound status.

As a component of the FPS game acoustic ecology, the soundscape, therefore, displays various characteristics betraying some of the relationships between players and between players and the game engine.[10] Furthermore, the use of elements of the conceptual framework aids in explaining these relationships. Thus, the existence of both volatile and predictable elements in the soundscape is a clue to player activity. The greater the volatility the more likely it is that such volatility is a result of increasing player action. Predictability is not only an effect of the game mode but also indicates a lower level of player activity (both on the part of the player whose soundscape is being analyzed and on the part of other players within exodiegetic hearing range of the single player). This is because environment sounds are sounded by the game engine with some degree of predictability — they may continually loop or may be sounded at regular intervals — and, being less likely to be masked by the sounds of frenetic player activity, they become more strongly identifiable in the soundscape. Other than environment sounds and some game status sounds, almost all sounds in the FPS game soundscape are predicated upon player actions and such actions have a lower degree of predictability particularly once the game play is well under way. The soundscapes of multiple players in the same game may have elements in common and these may be global or team-based sounds or character and interactable sounds depending upon the virtual distance between players. Furthermore, soundscapes may be analyzed using Schafer's terminology with the understanding that players, depending upon experience and context, will make their own decisions as to which sounds are to be foregrounded and which may be safely backgrounded.

9.1.2 The player in the acoustic ecology

Players themselves comprise (in the multiplayer game) a set of shared and varying experiences as they relate to sound and the FPS game world. These experiences contribute to the understanding to be gained from FPS game sound and, therefore, the players' level of participation in the game world. More experienced players are

---

9 As previously noted, realism FPS games such as *Urban Terror* typically have few such diegetic, symbolic auditory icons.

10 In the case of the multiplayer game, the game server plays a role too in coordinating the various game engines and by contributing instructions to them to play global sounds.

able to engage more quickly with other components of the acoustic ecology than those who are still feeling their way in the dark. The soundscape comprises all the audio samples which are used within any particular level and these will be processed further in various ways ranging from simple locational and depth processing (pan and amplitude) to any more sophisticated processing which the game audio engine may be capable of. All sounds have a set of affordances which will be prioritized by players dependent upon their experience of that particular game or of the sonic conventions of FPS games in general.

As an example using the Prelude of chapter 1, a novice player, upon first spawning in proximity to the church in a capture the flag, multiplayer configuration of the *Abbey* level of *Urban Terror*, will hear two specific (and predictable) sounds (the birdsong and the Bach organ fugue) and a range of sounds depending upon the actions of teammates spawning at that location at the same time. Such a player may be intrigued by the organ music, especially because it is a localized sound in the sound field which will move around the player as he turns his character around (unlike the globally dispersed and pervasive birdsong). At this point, the player may start to attend to this piece of music in any one of a number of ways. If he is musical, or is familiar with the works of Bach,[11] he may utilize the reduced and semantic listening modes in order to appreciate the musical qualities of the sound, a part of which process may be a recognition of the music. It may well be that causal listening is used as well which, when combined with prior experience, may lead to the surmise that this heavily reverberated church music (functioning in this case as both a choraplast and a topoplast) is likely to emanate from a nearby large, stone church.[12] He may notice, as he starts to move, that the music is shifting around his sound field and perhaps among his first actions in the game may be a switch to the navigational listening mode and the kinaesthetic following of the audio beacon to its source. If so, the confirmation that it does indeed issue from a large, stone church designed into the level is an aid to the construction of a navigable mental map of the level for future use.

The above scenario assumes the player makes certain choices, certain decisions as to which of the sonic affordances on offer to prioritize, some of which may be based on experience gained outside the game. However, as an experienced player familiar with the *Abbey* level, I myself (having initially followed the trajectory described above) now give a lower priority to the affordances offered by the organ music. I know what it is, what it signifies and, in part due to previously using the music as an audio beacon, I already have a mental map of the level layout and so have no need to explore in order to navigate anymore. It functions, for me, as a keynote sound rather than a signal sound of interest having, in the main, immersive, choraplastic and topoplastic properties indicating the spaces and locations I operate in as a player and aiding in my re-immersion into the game world. Nevertheless, as noted in chapter 8, I often cannot but help attend to the music, appreciating it musically and being drawn to the church locale of the level in order to kinaesthetically increase the intensity of the experience.

In order to be able to participate in, to be a component of the FPS game acoustic ecology, the player must be immersed within that ecology. As has previously been asserted (see chapters 6 and 7), this is accomplished by the soundscape both

---

11 An example of prior, ex-game experience being brought to bear.

12 The name of a level may also be a clue.

sensorially and perceptually, both in reality and in virtuality. The more immersive a game is the more appropriate it is to discuss the game world in terms of an ecology and, therefore, the greater the immersive function of the game's sounds (as attractors, retainers and connectors, for example) the more appropriate it is to describe the game's sounds as an acoustic ecology. For any organism to have an effect on other organisms, to have an effect upon its surroundings and, in turn, to be affected by those surroundings and other organisms, the organism must be immersed in those surroundings. The requirement to be immersed in the game to some extent, therefore, is a given when describing such a system and it may be stated that the greater the frequency and efficacy of immersive sounds, the greater the likelihood that the game's soundscapes and its players will form an acoustic ecology.

As part of its immersive sonic experience, *Urban Terror* provides a range of proprioceptive, kinediegetic sounds encouraging imaginative immersion through character identification (these may be classed as perceptual sureties which are consistent with the logic of the FPS game). Other forms of immersive experience, such as challenge-based immersion, are provided for by audio beacons, for example, or by the ability to locate an enemy sniper through the locational and depth parameters of a bullet's sound. Because all sounds, other than global game-status sounds, may have localization and depth cues applied, the player is encouraged to believe that he, through his character, is operating within the illusional 3-dimensional space depicted on screen. Furthermore, the more sophisticated use of sounds such as choraplasts, topoplasts and aionoplasts aids in imaginative immersion through the provision of virtual resonating spaces and paraspaces. Lastly, as shown in chapter 7, the player is able to have a significant impact upon the environment, the soundscape, through the production of kinediegetic sounds.

9.1.3 The relationship between soundscape and player

In *Urban Terror*, like most other run and gun FPS games, it is not possible for players to alter the visual, architectural environment of levels to any great extent if at all (that is, the visual structures of the level displayed on the screen may not be changed). Minor changes, depending on the configuration of the game, may include the opening and shutting of doors, the operation of lifts or platforms, the smashing of glass, the destruction of an object (such as a plate) or the picking up and discarding of various items. These are all trivial details;[13] in some cases they are only fleeting in their effect and the major structures of the architectural environment remain in place. Thus, an opened door is often automatically shut after a short time interval[14] and a smashed window may be automatically repaired after a minute or two in preparation for the next in-game hooligan; the buildings remain, the trees remain, islands do not move or disappear, train tracks stay in the same place and a grey painted wall remains a grey painted wall for the duration of the level.[15] Furthermore, the level's physical structures are more or less identical each time the level is played and by

---

[13] Which, nonetheless, offer other pleasures and reinforce the interactive nature of the game.

[14] This happens in *Quake III Arena* but doors usually remain open or shut in *Urban Terror*. In the former, it is the proximity of the player to the door which determines its state whereas in the latter, the player must actively open or shut the door with keyboard or mouse input.

[15] Discounting the temporary effects of pockmarked walls following a firefight or the in-game use of a spray-can of paint for tagging as is possible in *Counter-strike* (Valve Software, 1999).

whomever it is played.[16] Thus, the physical, visual environment remains substantially static throughout the playing of the level and, while this environment may affect the gameplay of players, players (beyond the nugatory and often temporary interventions described above) do not in turn affect this environment.[17] Any relation between environment and player is one-way only — for example, the physical structures of the game environment may direct players down certain paths or affect strategies of gameplay depending on where and how team bases or other objectives are situated.

The possibilities for player intervention in the soundscape of the FPS game are very different. In this case, much of the sound heard by any one player will be a result of his actions or the actions of other players (or, where sound is not heard, a result of lack of action — see chapter 7). While there may be environment sounds which are part of the level and global feedback sounds which may not be controlled by any player, there is a range of other player-initiated sounds to be heard. Depending upon the FPS game, these may be sounds of breathing, footsteps, doors opening or closing, the sound of a powerup being used, weapons fire, explosions, mechanical noises from vehicles, radio messages or sounds of pain to name but a few. Such sounds, whether they emanate from the player's character or from other characters within the game, are typically more frequent and more dense (especially at points of action) than level-specific or global feedback sounds outside a player's control. Thus, where a player's environmental intervention in the form of an opened door, for example, may have a small effect upon the architectural environment, the typical density and frequency of player-initiated sounds has a major effect upon the sonic environment. Unlike the physical, visual environment, which is set in place before the level starts and which is largely unchanging in its structures during gameplay, the sonic environment is created mainly by the actions of the game's characters and players (although drawn from the limited sonic palette of audio samples which is supplied with the game[18]) and is therefore not only created on the fly during the gameplay but has a different mix each time the game level is played.

This relationship between the characters' actions and the sonic environment is not just one-way though. The sonic environment, as has been argued in this work, also affects the behaviour of players (and therefore the actions of their characters) in the game and this is consistent with Schafer's view that humans in a natural ecology are shaped by the soundscape in which they live (pp.9—10). Truax (2001) defines the term *acoustic community* "as any soundscape in which acoustic information plays a pervasive role in the lives of the inhabitants [...] it is any system within which acoustic information is exchanged"[19] (p.66) and it is this pervasive role which the acoustic

---

16 Some FPS games, such as *Urban Terror*, may introduce minor structural changes depending on the game type configuration.

17 There are, of course, exceptions which, nevertheless, are rare enough not to disprove this general rule. In the *Urban Terror* level *Docks*, in a capture the flag configuration grenades lobbed at a particular wall will shatter it revealing a hidden passage. In this case, that section of wall behaves like a type of in-game door but remains open for the duration of the game.

18 With the exception of Voice over Internet radio communications which are an add-on to the FPS game.

19 This statement is a little confusing at first. Although Truax clearly states that the acoustic community is a soundscape that acts upon the inhabitants of that soundscape, the use of the word *community* implies, in common usage, an organization of organisms, particularly human, in which case such organisms (humans) are a component of that soundscape as acoustic community. However, although he does not elaborate further, I take his use of *community* to be that of a 'common character', a 'quality in common' or perhaps to be understood in the sense of, for example, 'community medicine' or 'community architecture'. I incline to the former because the correct usage of the latter sense would be 'community acoustics'.

community (the soundscape in the FPS game) plays that indicates the direction of the relationship. In this case, sound in the soundscape acts as the information nexus between players and between players and the game engine. In both cases this is especially the case where game events are unseen (having acousmatic sounds) and, as a significant example in *Urban Terror*, the probable reaction to the global game status sound indicating that a team's flag has been taken (and is on its way to the enemy's base) in a capture the flag configuration is indicative of this type of relationship.

That signal sounds are indicative of game action (much of which is predicated upon player action) is significant in the development of the notion that the FPS game soundscape is the environment of the FPS game acoustic ecology. An ecology details relations not only between the environment and its inhabitants but also details relations between those inhabitants. Therefore, if the game's soundscape is to be characterized as the acoustic ecology's environment, there must be relationships between players and the soundscape which are sonically based and which affect those players and that acoustic environment. This is the function of signal sounds which are carriers of the game's action and which are agents of change in the soundscape and of relational change between players. The game action of which they are indicative is brought about by player action and other players are invited to respond to the aural manifestations of such actions. Therefore, the sound of a shotgun firing in the vicinity of a player's character has several possible interpretations each of which the player may respond to and which, in so doing, will trigger further sounds inviting responses on the part of other players. For example, possible interpretations may be that there is an enemy firing at the player, that an enemy is nearby firing at another player or that a teammate is in the vicinity firing the shotgun (much depends on context and player experience not least upon the game configuration). Each of these possible interpretations invites different responses and each response, other than standing still, will trigger further sounds, such as more firing, the sounds of footsteps, of pain or radio messages appealing for aid and each of these will be heard by various players, potentially provoking responses on their part if they are treated as signal sounds.

Section 9.5 suggests that one possible way to view the relationships between the components of the FPS game acoustic ecology is that they are components of a dramatic performance. In this scenario, the players and the game engine are viewed as actors and the game engine also has the function of providing the script and the scenography. The script, in large part, consists of the game's diegetic, game-specific audio samples whereas the scenography, or *sonography*, is provided, in the main, by those level-specific audio samples which are environment sounds. The actors, within the confines of the game's objectives and conventions, semi-improvise the drama.[20] Böhme, discussing his concept of atmospheres in acoustic ecologies, states that atmospheres combine Production Aesthetics and Reception Aesthetics:

> Stage design is the paradigmatic example of this [Production Aesthetics] approach to atmospheres. On the other hand, however, atmospheres may also be experienced affectively, and one can only describe their characteristics insofar as one exposes oneself to their

[20] There are some similarities here to Terry Riley's seminal minimalist piece *In C* (State University Center of Creative and Performing Arts, 1968) although, in the FPS game, the order in which audio samples is delivered is less directed than the order of delivery of the musical fragments in the composition.

presence and experiences them as bodily sensations [Reception Aesthetics] (p.15).

Drawing a parallel to the performance scenario, it is the game engine that provides the Production Aesthetics and the players who, through Reception Aesthetics, experience the atmospheres of the FPS game acoustic ecology. Depending on the types of sound heard and their relationship to the image on screen, these atmospheres may comprise, for example, fear, exhilaration or threat. However, perhaps the main affective atmosphere experienced by players is one of immersion and participation, of being an integral part of an acoustic ecology (prior to other affective atmospheres which may be experienced such as the thrill of the chase, pride or excitement).

9.1.4 The relationship between player and player

As posited above, in a multiplayer FPS game, there are multiple acoustic ecologies, one for each player reflecting that player's point of audition within the game world. It has also been suggested that these acoustic ecologies are part of the larger acoustic ecology of the game world so here the relationships between the former and how they form part of the latter are discussed — in so doing, illuminating relationships between players. It should be mentioned here that the game's acoustic ecology is not a physical entity the environment of which may be recorded as with the individual acoustic ecologies. Rather, the concept refers to the virtual existence of an acoustic ecology in recognition of the fact that, although players will have their own unique sonic experiences, there is a common thread and shared components that, because they comprise soundscapes and players and relationships between these components, may also be termed an acoustic ecology. In this sense, it is similar to the concept of the game world; it is not possible to produce a snapshot of the game world as experienced by all players at one time and the concept of a game world is a shared, group experience which fleetingly exists in virtuality.[21]

Within any one player's acoustic ecology, a range of ideodiegetic character, interactable and feedback audio samples betray the presence and actions of other players who may experience those sounds as ideodiegetic too (especially if these players are the kinediegetic origins of the sounds). In *Urban Terror*, the first two classes comprise the sound of footsteps, breathing, bullet ricochets and bouncing grenades, for example. Feedback audio samples include team radio messages and game status signals (indicating, for example, the capture of flags). In the case of the former, such radio signals will form a part of the acoustic ecologies of all members of that team and, in the latter case, such global sounds are a part of each player's acoustic ecology. These are all examples of shared components of player's acoustic ecologies providing a common gateway to an understanding of and participation in the gameplay.

Telediegetic sounds, too, have a role to play in knitting together the multiplicity of

---

[21] Perhaps the closest attempt to capture such a snapshot would be a program-state snapshot of the computer code variables on the game server. With close scrutiny, this would enable, for example, an analysis of player locations, weapons being used and radio messages communicated and so may serve, to some extent, for an analysis of player relationships in the game world. Of course, game servers do this form of analysis as a matter of course, not just FPS games but also many other multiplayer games.

acoustic ecologies into one shared acoustic ecology. They provide a sense of history, that player actions have taken place in the past of the game time. Like the butterfly flapping its wings in the Amazon,[22] these sounds have consequence for the gameplay. In the one player's acoustic ecology, they are ideodiegetic and have immediate relevance for that player. His responses to that sound have potential consequence for other players who, telediegetically and in the future, experience a shadow of that sound as a remembrance of things past.[23] These sound events and the player's responsive actions provide a history of the individual gameplay and acoustic ecology of another player, in another location and at another time.

9.1.5 The technical production of the acoustic ecology

Chapter 4, suggested that it is possible to view the FPS game audio engine as a sonification system. This is a way to explain the technological genesis of the ecology; how the game engine provides a framework which technically enables the creation and maintenance of the acoustic ecology and which provides the means for the player to immerse himself within and participate in that ecology. Sonification, then, provides the building blocks of one component of the acoustic ecology, the soundscape, and, additionally, provides a relational framework for the participation of the other component, the player.

At its most basic level, sonification works by audification (0th order sonification); that is, it is a system that transforms the stored digital data of the audio samples into sound to be heard and interpreted by players in the FPS game. This is enabled through the use of computer code issuing instructions to digital audio cards (DACs) to perform digital to analog conversion of the audio samples and to transmit the electrical analog representation of the soundwave to the transducers (loudspeakers or headphones) that, finally, transduce the electrical signal into acoustic waveforms.[24] However, this does not explain the means by which a range of information is encoded into the audio signal. Audification simply explains the process of transferring non-acoustic data (the audio samples) into acoustic data (the acoustic waveform). Sonification techniques other than 0th are required to explain the higher-level encoding.

In all cases, sonification methods are predicated upon player action and presence. Sounds are only played in response to player action, or, at the very least, player existence within the game world, and this is the case not only for kinediegetic sounds but also for exodiegetic sounds. Thus, in the case of an environment sound (the Bach organ fugue in the *Urban Terror Abbey* level, for example), the ideodiegetic hearing of the sound requires the presence of the player.[25] The audio sample would simply not be sonified, would not be a part of the player's acoustic ecology, were he not present (in the form of his character) in that location. Furthermore, the action of

---

[22] The apocryphal basis for Chaos Theory.

[23] Not that I am suggesting players should take time out of gameplay to ponderously reflect on the memories engendered by a nibble on this Proustian biscuit.

[24] Depending on the hardware the player is using, in some cases, it is not the DAC that performs the digital to analog conversion but the amplication system which receives a digital signal direct from the DAC. Furthermore, the system described here is a simplification and there will be other processes undertaken by either the game code or by the DAC such as the mixing down of multiple audio channels into stereo or 5.1 sound for surround sound speakers, for example.

[25] This is similar to Böhme's suggestion above that atmospheres in the acoustic ecology require a discerning subject to be present.

the player too has an affect upon this sonification and he may exercise a kinaesthetic control over that sound by moving his character in relation to the position of the sound source (here, the church that is part of the level). In this case, by moving a certain distance away from the church, the sound is attenuated until the point at which it ceases to play. In technical terms, the game engine tracks the character's position within the virtual space of the game world in relation to the sound source and decreases the volume of the audio sample until it is stopped altogether. By the simple act of fading the volume of the sound (and this works in reverse too), this form of sonification provides a relational framework for the player to begin to contextualize himself within the resonating spaces and paraspaces of the acoustic ecology.

As regards kinediegetic sounds, as the term implies, these audio samples are also only sonified through player actions. Thus character audio samples and interactable audio samples are sonified in response to the player's haptic input. Within that player's acoustic ecology, those ideodiegetic types of audio samples which are exodiegetic typically betray the actions and presence of other players (or the emulated actions of players in games where bots are used) and thereby provide a part of the relational framework of the acoustic ecology which enables the player to contextualize himself in relation to other characters within the game world. Game status sounds, such as *Urban Terror*'s audio samples signalling the flag status in capture the flag games, are also indications of player activity.

Concerning the real-time processing of audio samples with reverberation cues (as found in more sophisticated game engines but not in *Urban Terror*), sonification provides another level of information translation. Here, the sonification which takes place is one that translates the material and spatial properties of the visual spaces and paraspaces of the game into resonating spaces and acoustic paraspaces. And, once more, this sonification increases the immersiveness of the gameplay experience by furnishing the player's acoustic ecology with a range of perceptual sureties and surprises to do with the spatial parameters of the game world. Less sophisticated game engines, such as that found in *Urban Terror*, are able to do this to a limited extent through the provision of pre-reverberated audio samples which possess a general purpose outdoor or indoor choraplastic function.

This completes the description of the FPS game acoustic ecology. That description encompassed the components of the ecology, the relationships between these components and an explanation of the technical genesis of the ecology such that it is now possible to state that the sound heard when playing an FPS game may be treated as an acoustic ecology. As a model of the relationship between player and sound in the FPS game, it is more comprehensive than those models of digital games discussed in chapter 2. However, what has been described above is just one player's acoustic ecology and from one player's perspective. The section that follows argues that a multiplayer FPS game comprises multiple acoustic ecologies all contained within the game's virtual acoustic ecology and, to do this, makes use of autopoietic theory.

9.2 The acoustic ecology as the phenomenological domain of an autopoietic system

In a multiplayer FPS game, there is not just one acoustic ecology but several and, furthermore, these operate within a virtual acoustic ecology. In ecology (that is, the biological study), it is possible to study the ecology of, for example, a sandy riverbed, the ecology of a nearby kopje and the ecology of the termite mound and these may all be seen as inter-related components of the ecology of the Kalahari scrub. So it is with the multiplayer FPS game where each player and his soundscape form a unique entity which may be studied (as I have done here) as an acoustic ecology by itself but where each acoustic ecology forms part of a larger whole. Like the real-world ecology, the components of such individual ecologies are not necessarily fully aware of other acoustic ecologies (the queen termite, while fulfilling her reproductive function, operates solely within her own) but they are affected by them and, indeed, may have aspects in common. Accordingly, this section describes the operation of the game's virtual acoustic ecology, comprising individual acoustic ecologies, and frames this idea in autopoietic theory.

The operation and maintenance of the virtual acoustic ecology may be explained through applying the principles of autopoiesis to an expanded and simplified version of the model in chapter 8. In this model, I suggest that each player's acoustic ecology may be viewed as the phenomenological domain of an autopoietic system (comprising the allopoietic components of player, soundscape and game engine) and that this autopoietic system itself is an allopoietic component of the autopoietic system that is the virtual acoustic ecology of the networked, multiplayer FPS game.[26]

An autopoietic system is a homeostatic organization devoted to the maintenance of that organization. External information is viewed as a perturbation to which the autopoietic system responds by compensatory processes of transformation (production and destruction of its components) to further the goal of maintaining its organization as an autopoietic system. By this definition (and focussing on a single player's FPS game acoustic ecology), the autopoietic system comprises the FPS game engine (containing the computer code and the game's audio samples), the player and soundscape and the system's phenomenological domain is the acoustic ecology. The purpose of the system is the preservation of its organization as an autopoietic system which entails the maintenance of its phenomenological domain, the game's acoustic ecology. If that domain is defined in terms of its ability to indicate spaces, places and times in addition to indications of player activity, then the transformations of the system's component sounds are compensations for the perturbations in the networked, multiplayer FPS game. These compensations are for the maintenance of the structures of the acoustic ecology (these structures may undergo transformations as long as they remain structures defining the acoustic ecology). If this fails, if the player is no longer able to perceive the acoustic ecology of the game, then the acoustic ecology no longer exists as a phenomenological domain and the autopoietic system has failed in its purpose and is, therefore, no

[26] While an extended view of FPS gameplay (a game level in progress) as an autopioetic system is beyond the scope of this thesis, it may be a task for future research. Additionally, Puterbaugh (1999) has defined the term *sonopoiesis*: "Sonopoietic space is the space of listening that we create through the act of listening to sound". However, this is insufficient for an understanding of the FPS game acoustic ecology as the phenomenological domain of an autopoietic system as it does not account for the player's haptic input and ability to trigger sounds. It is a listening space that is less actively created by the listener, and therefore more akin to cinema, than the actively-created listening and participatory spaces found in FPS games.

longer autopoietic.

As new players join the game, the autopoietic system that is the virtual acoustic ecology responds to this external information by undergoing transformation (for example, the inclusion of a new allopoietic component that is the new acoustic ecology) as a way of compensating for this perturbance. These transformations ripple through the system as perturbations themselves, impinging first, and with greatest effect, on those players closest (in the game world) to the new player with the result that the autopoietic acoustic ecologies of these players themselves undergo compensatory transformations which are manifested as new sounds or the stopping of sounds (the production and destruction of components). Additionally, it may well be that the transformations of the virtual acoustic ecology form a part in enabling the new player's immersion in the system because the process of compensation is one of inclusion.[27] That is, the compensation takes the form of the inclusion of a new allopoietic component (the player and all his actions) and that this inclusion immerses the player in his acoustic ecology and the virtual acoustic ecology and, as a result, immerses him in the game world. Furthermore, it may also be postulated that if the player, an allopoietic component of the acoustic ecology, is viewed in autopoietic terms, then the compensatory responses of the player to perturbations arising in the soundscape lead to immersion in the game's virtual acoustic ecology. This process of poiesis (the creation of the player's soundscape) and immersion in the game's virtual acoustic ecology is demonstrated in *Figure 9.4.*

---

[27] In this case, it may be suggested that the player, upon becoming a willing 'immersee', sacrifices an autopoietic existence outside the acoustic ecology to become an allopoietic component of the autopoietic ecology and perhaps this in itself is a guarantee of immersion in the circularity of the autopoietic system.

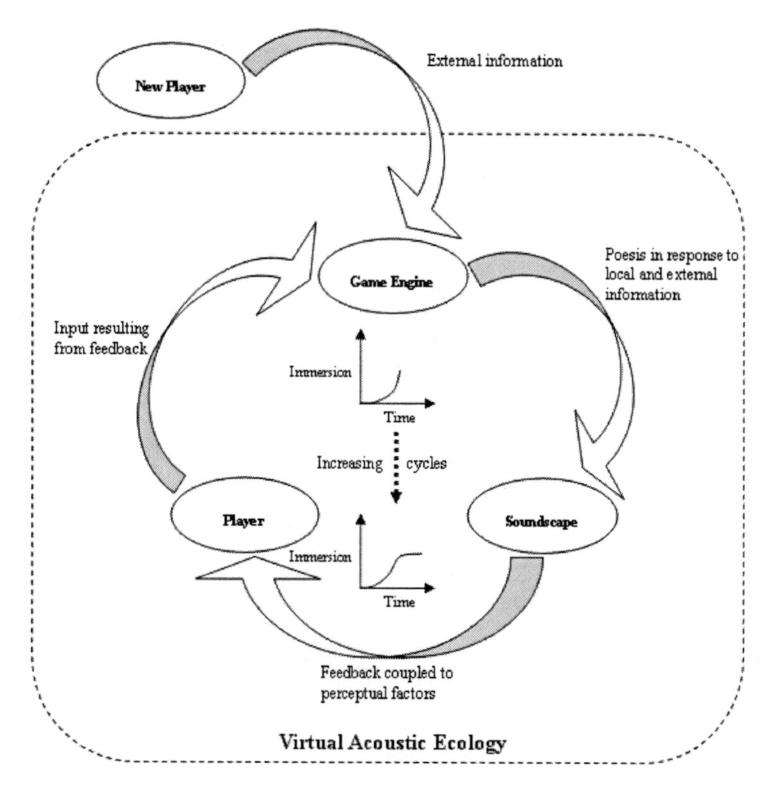

Figure 9.4. Autopoietic processes and player immersion in the FPS game virtual acoustic ecology.

In the diagram of *Figure 9.4*, the cyclical nature of the autopoietic system that is the the game's virtual acoustic ecology is demonstrated through a chain passing from player to game engine to soundscape back to player. Poiesis (sonification) is the process by which the soundscape is created in response to player input both internal and external. The soundscape provides feedback to the player that is filtered through a range of perceptual factors as listed in the model shown in chapter 8. This in turn provokes responses from the player which, through haptic input, are fed back to the game engine continuing the cycle. The initial cycle creates the soundscape, which then becomes a component (with the player) of the individual acoustic ecology, and the following cycles work towards a process of equilibrium in the soundscape (through the process of destruction and construction — that is, the sounding and silencing of audio samples) — a self-organized equilibrium that is characteristic of autopoietic systems. Furthermore, as this equilibrium is reached, the player's immersion in the game's virtual acoustic ecology is increased to its maximum because the ecology's compensatory transformations involve the inclusion of external information (the player). It is likely that greater player experience (a perceptual factor influencing the meaning given to sounds in the FPS game) will decrease the number of cycles required for full immersion-equilibrium.

Perturbations in one player's acoustic ecology are derived from the compensatory processes occurring in the game's virtual acoustic ecology and these processes result not only from the inclusion of new players, as noted above, but also from the ongoing actions of all participating players in the game. Here, telediegesis proves useful in explaining how this occurs. Perturbations in the system may be conceived of as ripples expanding throughout the system[28] but, as they expand, lessening in intensity. This is a process that takes place over time and, to explain it as an instance of telediegesis, the discussion returns to a part of the scenario described in the Prelude to chapter 1 which was taken from *Urban Terror*. Here, the hero is about to save his team's flag-carrier from the attack of an enemy soldier. At this point in time, there are at least three players in the scene. This conjunction has not happened by chance — sound has played a key role. In the case of the enemy, she has been following the flag-carrier through the twisting passageway's of the abbey's cloisters by tracking the sound of footsteps and gunfire (navigational listening). Similarly, the player has been drawn to the unfolding drama by the sounds of distant battle (again navigational listening). Thus the conjunction is a result of players' responses to exodiegetic sound which, in the initial phases of the chase, would not have been heard by the opposing player. This is telediegesis; the reaction by the enemy to the sounds of the flag-carrier having later consequence for the player (and his team) such that he comes face-to-face with that enemy and is able to save his team-member. Thus, telediegesis is a perturbation rippling through the virtual acoustic ecology from the enemy's acoustic ecology to the player's and vice-versa. Both these autopoietic systems compensate for these telediegetic perturbations through the manifestation of new, common sounds (the production of new components) thereby contributing to the maintenance of the phenomenological domains that are the players' acoustic ecologies and the game's virtual acoustic ecology.

The suggestion above that the game's virtual acoustic ecology is a phenomenological domain needs to be treated with care. As Maturana and Varela (1980) state, "autopoiesis generates a phenomenological domain, this is cognition" (p.123). Thus, the existence of the acoustic ecology presupposes cognition on the part of the autopoietic system where cognition includes the ability to acquire, store, retrieve and use knowledge. The FPS game engine (and associated game components such as computer code and computer memory, for example), in its capacity to acquire and store game information, to retrieve and use that information for the maintenance of its organization and 'phenomenological domain', is therefore, by this definition, cognate. The logical autopoietic conclusion then, is that the FPS game is living because it is autopoietic and autopoiesis is the sole requirement for life (according to autopoietic theory, "autopoiesis is neccessary and sufficient to characterize the organization of living systems" (Maturana & Varela, 1980, p.82)).[29] It is not the purpose of this work to argue that the FPS game's acoustic ecology is a sentient being nor does it.[30] But it is an interesting topic for future debate because here, it has been demonstrated that there are many aspects of the virtual acoustic ecology which may be described and explained through autopoietic theory such that it becomes possible to tentatively suggest that this ecology is an autopoietic system or, at least, a non-sentient Self-Organizing System.

---

28 Similar to the wave model of sound in acoustics.

29 Look out for FPS game rights advocates in the near future.

30 Particularly as the game engine is not able to sense sound or, indeed, any of the acoustic phenomenological domain to which it contributes.

There is a stronger argument to make that each individual player's acoustic ecology is a phenomenological domain arising out of a cognate autopoietic system because such a system includes the sentient player. Furthermore, the player is fully aware of all elements of his acoustic ecology and thus it may justly be described as his (acoustic) phenomenological domain. However, in the virtual acoustic ecology, though it contains players, and though each player's acoustic ecology may share common sounds with others, each player is unaware of the totality of another's acoustic ecology and, furthermore, unaware of the totality of the virtual acoustic ecology. Swinging the argument the other way, though, throughout this work it has been argued that the majority of sounds in the FPS game arise out of, and are, therefore, evidence of player actions and, thus, player presence. The virtual acoustic ecology may be said to be maintained almost entirely by the active participation of players reacting to other allopoietic components of the system. The virtual acoustic ecology may, therefore, be seen as a phenomenological domain produced by multiple, sentient players each of whom perceive just a portion of it. If a termite colony ecology may be seen as an autopoietic system,[31] there are, similarly, components of that system (the queen, for example) who, it may be argued, only perceive a portion of it. However, the FPS game (acoustic ecology) as life-form, or at least emergent system (see Johnson, 2001, for example), is a subject for future debate.

The concept of the multiplayer FPS game virtual acoustic ecology as autopoietic system is demonstrated in the diagram of *Figure 9.5*.

---

31 I am unaware of any writing explicitly describing termite colonies as autopoietic systems. However, termite colonies are often used as examples of such systems and other similar Self-Organizing Systems (for example Georgantzas, n.d.) and, having undertaken an A-level biology project on termite colonies in Botswana, I have some trivial knowledge of the construction of their mounds and organization of their colonies.

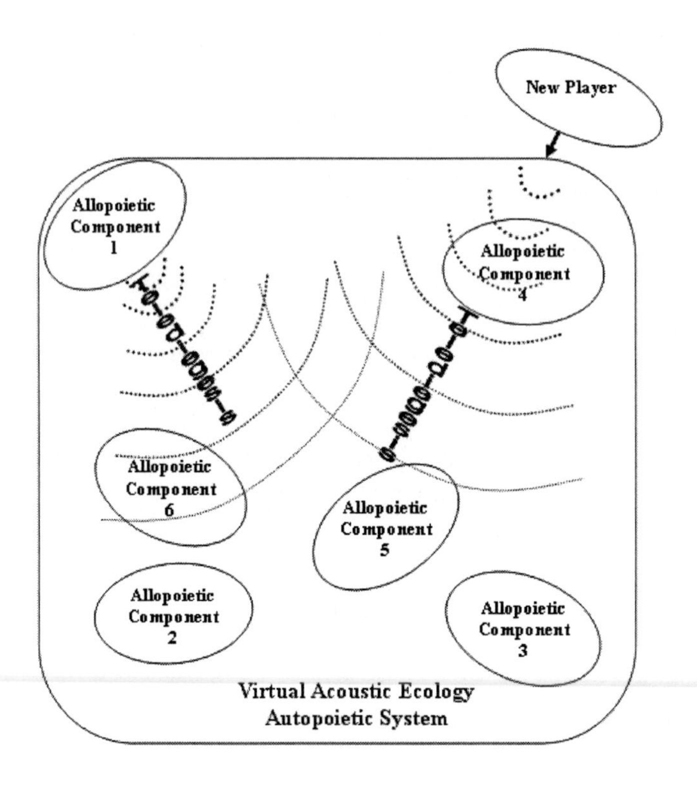

Figure 9.5. The multiplayer FPS game's virtual acoustic ecology as autopoietic system.

Figure 9.5 is a schematic of a multiplayer FPS game's virtual acoustic ecology in which five allopoietic components (comprising player, game engine and soundscape) are positioned. Telediegesis is used to explain the ripple of perturbations as a new player (external information) joins the game having the strongest effect upon allopoietic components within closer virtual hearing distance and a weaker effect upon those more distant. Within the virtual acoustic ecology, allopoietic components send ripples of perturbations[32] to other allopoietic components which, when viewed as autopoietic systems, respond to this external information with compensatory transformations of their phenomenological domains (their acoustic ecologies) thus sending forth further perturbations. The allopoietic components move throughout the space of the virtual acoustic ecology as their player components move throughout the game world, jostling with other allopoietic components and thereby deriving new exodiegetic sounds for their acoustic ecologies and contributing exodiegetic sounds to other acoustic ecologies.

---

32 For diagrammatic simplicity, only two such ripples are shown.

9.3 Testing the conceptual framework and model

The Prelude to chapter 1, provided a narrative description of one player's experience of a capture the flag scenario in *Urban Terror.* It is returned to here in order to demonstrate how elements of the FPS game acoustic ecology model presented at the end of chapter 8 may be used to analyze and conceptualize the sonic aspects of the FPS multiplayer game as dramatic performance, part directed, part improvised. While narrative may be found in FPS gameplay, to analyze FPS games and, in particular, the game's sound from narratological premises not only risks overlooking the genre's *raison d'être* but also, ultimately, provides little insight into the game much less insight into the complex web of relationships found in the FPS game acoustic ecology. It is much more worthwhile to view the FPS game acoustic ecology as a dramatic performance because this provides important insights into the nature of the relationships which exist within that ecology.[33] This exercise helps to demonstrate the robustness and versatility of the conceptual framework and model such that they may be used, or aspects of them may be used, for the analysis of FPS game sound from a particular conceptual perspective. This is what this section sets out to do.

Many game theoreticians apply the theories of narratology to digital games, (Carr, 2006; Juul, 2001; Liang & Tan, 2002 for example), sometimes, it seems, in an almost indiscriminate manner regardless of game genre. That being said, there are genres of digital games which prove amenable to narrative analyses; RPGs and their like are a good example as shown by Carr (2006).[34] Other theoreticians point to weaknesses in narratology as applied to digital games. Ryan (2001) for example, or Grodal (2003) who suggests that "a purely linguistic model may seriously impede descriptions of those media like video games that rely on a series of nonverbal skills" (p.133). Johnson states that part of the problem of understanding games is the application of narrative theories:

> [O]ne of the problems we have in understanding games is that we see them as being driven by their narratives. In fact, I think the narratives tend to be a vestigial part of games that has been carried over from earlier forms. When people play games, they aren't playing them for the story. They aren't playing them for a narrative arc of any kind. In fact, if you're looking for an analogy, I would say that game design is closer to architecture than it is to novel writing. The designers do create resistances to certain types of behavior and encourage other types of behavior within the space, but first and foremost, they're creating a space that can be explored in multiple ways (quoted in Wasik, 2006, p.36).

To extend his analogy, it may be suggested that the architectural space he mentions is, in the case of FPS games at least, the theatrical stage and its associated props and scenography representing a 3-dimensional space in which the player is situated.

---

[33] Future research may usefully examine the entirety of FPS gameplay from such a perspective.

[34] This may also be illustrated by the short back stories the developers give to FPS games and the almost complete ignoring of any narrative elements in FPS game reviews (see the *Quake III Arena Review* (n.d.) for an example that does not mention the back story or any narrative elements at all). Some fans of a more narrative bent have expanded these back stories into novella form (*Fan fiction inspired by Halo* (n.d.) for example) building upon elements hinted at in the game manual or gleaned from the game environment and gameplay but these are extra to the game, not a part of the official release and, being textual and literary, are a subject not relevant to this thesis.

Because the work has previously described resonating spaces in which the player is physically and virtually immersed, this analogy may be extended to the FPS game acoustic ecology.

Rather than attempting to analyze a visceral, action-based game such as the FPS game as narrative, it is a more relevant exercise to discuss FPS games as theatre where the player is a character performing on the stage of the game world. Although he never uses the word drama or suggests that (some) games may be viewed as performance, this, I would propose, is one meaning that may be taken from Malaby's (2006) statement that "making a game is not, as some narratologists would have it, about making a "story", it is about creating the complex, implicit, contingent conditions wherein the texture of engaged human experience can happen" (p.9). This is of relevance to the hypothesis; viewing the gameplay of the FPS game as a performance is one particular method of analyzing the role of the player in the game surrounded by other actors, paying attention to sound cues and contributing in kind to the acoustic ecology. As Malaby further states: "Games [...] are about contriving and calibrating multiple contingencies to produce a mix of predictable and unpredictable outcomes" (p.9).[35] Laurel makes the point that, rather than discussing human-computer interfaces (and here she includes digital games) from the point of view of narrative, it is more worthwhile to assess them as theatre. She applies Aristotle's four causes (formal, material, efficient and end cause) to theatre in an effort to explain computer software and interfaces (Laurel, 1993, pp.40—48). This is an interesting point and particularly relevant to FPS games where the player is a performer in the drama being enacted within, and that is an enabler of, the game and it is a view that is reinforced by Murray (2000) who, discussing the equation between real-world movements (haptic input such as the keyboard and mouse) and virtual world movements in video games, states "this is not a passive board game but a live-action stage" (p.108).[36]

Although it would be of interest to analyze the FPS game in its entirety as a dramatic performance, here I wish to use the fundamentals of performance and drama to analyze solely the game's acoustic ecology. In doing so, this will, in part, elucidate aspects of immersion within and participation in that ecology as they are effected and affected by sonic relationships between players and between players and the game engine.[37] Particularly in a multiplayer game, the gameplay may be viewed as a social phenomenon, but this view and its implications may also be applied to single-player FPS gameplay if, with not too much a stretch of the imagination, AI-enabled bots may be seen as (simplistic) social participants.[38]

What are the dramatic components of the FPS game acoustic ecology and by what

---

[35] It may be possible to marry some of the ideas of narratologists to an ecological view of the game forming a theory of the *narrative ecology* of digital games. This is, after all, a fair description of games as theatre making use of narrative (story and plot), contrived yet calibrated multiple contingencies in the form of players, bots and game settings, for example, and both predictable and unpredictable outcomes (different scores and player rankings resulting from a variety of factors including player skill, chance and network latency, for example). But that is a matter for another thesis.

[36] Although, as noted in chapter 1, Eskelinen (2001) warns against the use of the principles of Aristotelian drama with regard to digital games.

[37] As noted above, here, 'game engine' is a conceptual term that includes individual player's game engines and the game server coordinating the game.

[38] However, as bots, whilst producing sounds, do not respond to sound, this analysis concentrates on the multiplayer FPS game acoustic ecology. The analysis of this social phenomenon as drama has parallels to actor-network theory where such phenomena are "ordered networks of heterogeneous materials [and where] the social is nothing other than patterned networks of heterogeneous materials" (Law, 1992, p.2).

processes do they relate to each other? The actors are the game engine and the player(s). Breaking these components down further provides clues as to their relationships to each other, how they respond to each other and how they interact with each other. The player (in the acoustic ecology) becomes the character expressed through a range of proprioceptively- and exteroceptively-sounded kinediegetic audio samples which are set in play by the game engine in response to player actions. These kinediegetic sounds signal the player's presence on the stage of the acoustic ecology and provide the cues for other (human) *dramatis personæ* to respond in a manner similar to the way in which the actor's voice is her sound presence on stage (Berry, 1987, p.16). The game engine is the acoustic ecology's stage, its superstructure, and, furthermore, not only provides a range of scenographic environment sounds (the *sonography*) and other sonic props but also furnishes the game script (character and interactable audio samples) within which there is much leeway for improvisation in terms of the sonic plot as the story is followed.[39] This is achieved through the process of sonification as described above. However, in its guise of responsive computer code, the game engine is also an actor in the acoustic ecology. It may sometimes take a minor role as prompter — in its provision of sonic cues such as audio beacons for example — or it may take a more active role on the stage by providing sounds such as game status signals to which players respond and which, in most cases, will have an effect upon the plot construction. Furthermore, the more sophisticated FPS game engine is able to recast the player's sound through real-time DSP that places the player's character in different spaces, locations and times. The less sophisticated FPS game engine (such as *Urban Terror*) is able to do this to a certain extent through the use of pre-reverberated audio samples and the attenuation and amplification of environment sounds. Audio samples functioning as choraplasts, topoplasts, aionoplasts and chronoplasts serve this purpose.

The acoustic ecology of the FPS game, then, is contrived with the aid of heterogeneous components which comprise players, game engine and audio samples and which, in their use and relationship to each other, provide the multiple contingencies of the dramatic performance. The use of these components leads to the contrived contingencies which bring about both predictable and unpredictable outcomes in the FPS game acoustic ecology and which mark it as a dramatic performance. The actors, the players and the game engine, create, through the production of sound, the sonic drama that is the acoustic ecology. This production of sound is, in the case of players, a result of player actions which may be in response to the hearing of other players' sounds (their sonic cues) — there is a certain amount of free will in the player's sounding of sounds hence the improvisatory nature of the performance and the aforementioned mix of predictability and volatility in each player's soundscape. In the case of the game engine, the production of sound, where it is not a part of the sonography, is in response not to the hearing of sounds produced by other actors in the drama, but is in response to the actions of those actors — the game engine cannot hear sound, it can only play sound.

In most cases for a staged drama, a play performed in a theatre, the actor follows other actors' speech in order to participate in and contribute to the drama. In order to follow the gameplay, to work in tandem with teammates and hunt out the enemy and to situate himself in a particular point in the environment, the FPS player usually

---

39 For a capture the flag game, this story is likely to be: two teams compete to capture the most enemy flags by the close of play. Within the constraints of this sparse narrative story, multiple plots may occur and, in multiplayer games, may occur simultaneously.

relies not on dialogue/speech but on non-speech sound cues from the environment, whether those cues are other characters' kinediegetic sounds or are emitted by the environment itself (these sounds may be wind, birds, waves, animals, machinery — environment sounds in other words). In some cases, such sound works with vision by the process of synchresis for that part of the game environment that is visible (c.90°–120° on the monitor in front of the player and depending on the player's distance from the monitor) but, for the world of the game that is not visible, sound works alone.

Returning to the scenario played out in the Prelude to chapter 1, it is now possible to identify the dramatic components of this performance using aspects of the conceptual framework devised throughout this work. The script consists of audio samples to be used according to the rules and conventions of the game as directed by the game engine. The actors are the players. The game engine itself takes a leading role. Importantly, though, the game engine also provides the stage of the performance (the means by which it can take place) and its sonography, comprising environment sounds, and this provides the basis of the various acoustic spaces in which the player is immersed and performs. Through haptic input, the player delivers his lines as snippets of sound and such kinediegetic sounds are representative of the player's actions. These, then, are the exodiegetic cues for the use of other players within hearing range. Depending upon which cues are heard, these other players select and deliver their own kinediegetic sounds. The choice of antiphonal response depends upon the player's use of the functions of sound, as interpreted by the context and the player's experience, and this affects the perception of all ideodiegetic sounds for that player. Ultimately, this leads to the player's own appropriate contribution to the performance.

In the Prelude, therefore, all players' scripts comprise their character sounds, interactable sounds and feedback sounds and these are all kinediegetically controllable. The hero delivers lines consisting of footsteps, grenade explosions and shotgun blasts which are heard as exodiegetic by other players. These other players respond in like manner with kinediegetic lines of their own or follow their own performative arc. The player's choice to move towards the church is prompted (assuming this is a premiere performance) by the game engine providing an audio beacon (the Bach organ fugue) which he attends to in the navigational listening mode. It is the game engine, in its guise of actor, that sounds the alarm (heard by all players) when the enemy's flag is taken. The player's response to this feedback sound is, in part, dictated by his experience of the conventions of the game and by the context. In this case, as a member of the attacking team, he moves to support the flag-carrier. In the process, he contributes a plethora of sounds to the performance indicating his level of activity — footsteps, grenade explosions, shotgun blasts — all the while attending to (and occasionally responding to) the affordances offered by the sonic cues delivered by other players.

There are two important points to note here. First, there is an element of improvisation in the player's performance. The 'story' of the level may dictate that two teams battle for each other's flags, indicating that certain global sounds, at least, will form part of the performance, but it is his choice to follow the organ music and it is his choice which weapons to use. Furthermore, such choices may well be different the next time the acoustic ecology is performed. For example, at the next performance, it may no longer be necessary to listen to the organ music in the navigational listening mode because the previous performance has led to the construction of a mental map of that part of the level — the church is to the right of the spawn point. Secondly, were the scenario of the Prelude to be given in script form to the players prior to the performance, each script, while exhibiting similarities, would be substantially different.

The similarities are illustrated by the common sounds shown in the model of chapter 8. These are ideodiegetic sounds, heard by all or a group of players, such as those generated by player activity within hearing range of the script's character or global feedback sounds performed by the game engine. However, in the multiplayer game, there are multiple acoustic ecologies unique to each player but all forming part of the game's virtual acoustic ecology. These are multiple performances and, in addition to the common sounds detailed above, the thread holding them together is telediegesis. Telediegesis, though, operates over time — it is a delayed effect moving from one player's acoustic ecology to one or more other acoustic ecologies. In dramatic parlance, telediegesis is akin to the dramatic aside — sounds performed in one part of the virtual acoustic ecology but unheard in other parts. However, in a form of dramatic irony, such hidden sounds and their accompanying player actions return later to haunt the unhearing player. An example from the Prelude is the series of events leading from the taking of the enemy's flag to the player's rescue of the flag-carrier. In this case, the enemy, following the sounds of his footsteps and gunfire through the twisting abbey cloisters (in the absence of a clear field of view), have driven the flag-carrier towards our hero who is making his way from the church. Were this hunting down not to happen, the player's performance would be very different.

These similarities and differences arise out of Malaby's contrived and calibrated multiple contingencies. Apart from the rules and conventions of the game, this contrivance and calibration takes place in the area of game sound design where the contingencies are the meaning designed into each sound affected by the context in which it sounds. What Malaby does not explicitly take account of, though, is the effect of player experience and knowledge and the improvisatory nature of the players' actions. The player has a certain choice over which sounds to use or whether (or how) to react to other heard sounds and this adds another layer of unpredictability to the performance.

Using parts of the model of the acoustic ecology, it is possible to produce a schematic of the FPS game acoustic ecology as dramatic performance as shown in *Figure 9.6*.

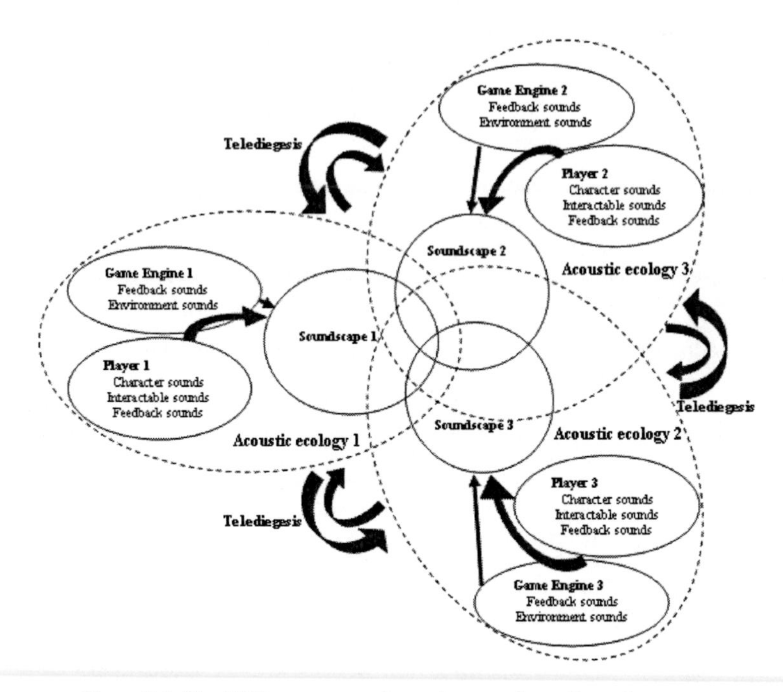

Figure 9.6. The FPS game acoustic ecology as dramatic performance.

Figure 9.6 is a simplified schematic of a multiplayer FPS game acoustic ecology showing just the contributions of three players and their game engines. It demonstrates commonality of some sounds between two players or between all, the processes of telediegesis and the basic script each actor has to work from. What it does not show, for simplicity's sake, is the contribution of game status feedback sounds on the part of the game server.

This, then, is one possible conceptual perspective on the FPS game sounds and the player's relationship to them — that of the acoustic ecology as dramatic performance — which, for its analysis, uses parts of the model of the FPS game acoustic ecology developed throughout this work. The lines delivered by actors are drawn from the scripted audio samples, the game engine, through the process of sonification, provides the stage and sonography and interjects with lines of its own while players use sonic cues from the game engine and from other players to participate in and to carry forward this performed drama. The player's act of creative participation within the loose confines of the FPS story makes use of perceptual (sonic) cues and surprises (given the improvisatory plotting of the story) and this leads to the immersion within the superstructure of the game's acoustic ecology, the setting foot upon the stage, and within which the player delivers his own sonic cues thereby contributing to the acoustic ecology himself.

9.4 Conclusion

This chapter has demonstrated that the hypothesis that the sound of the FPS game is an acoustic ecology in which the player is an integral and contributing component may be used to conceptualize the relationships between soundscape and player in the FPS game and has used *Urban Terror* as an illustrative example. The components of such an acoustic ecology are the player and the soundscape and the relationship between them is based on sound. Players participate in the construction and maintenance of the acoustic ecology by triggering sounds and the soundscape affords information in the form of sound for players to engage with and to potentially respond to. These are complex relationships and, while there are behaviours in the acoustic ecology which may be predicted (especially if the game configuration is known — free for all or capture the flag, for example), it is, for the most part, a volatile, shifting ecology. In this, although its sonic palette may be smaller, it emulates real-world ecologies.

It has also been demonstrated that there is not just one acoustic ecology in a multiplayer FPS game but several, with each player participating in an ecology unique to his experience but sharing some similarities with others' acoustic ecologies and that these form part of the game's virtual acoustic ecology. The processes by which the game's virtual acoustic ecology is created and maintained may be explained through autopoietic theory. Autopoiesis also helps to account for the phenomenon of player immersion in the acoustic ecology by suggesting that such immersion results from compensatory and inclusive responses on the part of the ecology.

Ultimately, the robustness and versatility of the FPS game acoustic ecology model has been demonstrated by using parts of it to describe that acoustic ecology as dramatic performance. In this use of the model, the players and the game engine are the actors and the script consists of the game's audio samples which are delivered in a part-directed and part-improvised manner. The game engine has the additional role of providing the framework in which the performance takes place and this is accounted for by the sonification of environment audio samples creating the performance's sonic stage and sonography.

The final concluding chapter summarizes the major points of the work, assesses the strengths and weakness of the theoretical model and its contribution to Games Studies and outlines potential foci of future research building upon the conceptual framework and FPS game acoustic ecology model presented here.

Chapter 10

Conclusion

This work has presented the hypothesis that the combination of player(s) and sound in the FPS run and gun game may be conceptualized, and the relationships between the two understood, as an acoustic ecology. Due to the lack of academic writings on digital game sound, let alone FPS sound, a multidisciplinary approach was taken in order to answer the series of questions posed in chapter 1. Therefore, chapters 2 to 7 comprised a critical review of the range of disciplines chosen, culminating in a conceptual framework and diagrammatic model of the FPS game acoustic ecology (chapter 8) and a development of the hypothesis in chapter 9. This chapter briefly discusses the strengths and weaknesses of the framework and model, their contribution to some of the disciplines researched (in particular Games Studies) and identifies areas for future research.

The research assessed in the work led to the identification of a number of gaps in current digital game sound theory. Of particular importance was the lack of any clear definition and understanding of the role of the player and his relationship to the soundscape. In order to tackle this, a multidisciplinary approach was adopted that ultimately led to a more comprehensive and flexible model and conceptualization of digital game sound than is to be found in the literature surveyed in chapter 2. The notion of an acoustic ecology, presupposing a relationship between listener and sound, has been a significant contribution to digital game sound theory. Thus, the work has been able to establish the player and the soundscape as the two major components of the ecology and has adapted or newly defined a range of concepts explaining the relationship between them. Importantly, through the use of the concept of telediegesis and autopoietic theory, the work has been able to make a start in explaining the sonically-based relationships that exist between players in a multiplayer FPS game.

The hypothesis' elucidation of the relationships listed above is an important addition to Game Studies and other research fields such as acoustic ecology and virtual environment design. This is particularly the case for questions concerning player or user immersion in digital games and virtual environments where the adaptation of existing theories on space and immersion and the definition of new concepts may prove to be of use in understanding the immersive and participatory experience. It is to be hoped, too, that the conceptual framework and model of the FPS game acoustic ecology will benefit (game) sound designers in their understanding of how sound functions in 3-dimensional virtual environments. In the case of sound design for FPS games, of key importance is the centrality of sound to immersion in the game world. It is sound, and the player's relationship to sound and consequent immersion within and participation in the acoustic ecology, that is key to immersion within and participation in the FPS game world. It is a firm belief of mine (now supported by this research) that sound should not be 'bolted on' to digital games during the design and development process but that it should be placed on a par with programming, modeling and gameplay design and that this will lead to more immersive digital game experiences.

Chapter 1 raised four questions about sound in the FPS game that were suggested by the hypothesis. It is now possible to supply answers to these questions that, if not necessarily comprehensive, do provide a basis for greater understanding about that sound and which help to guide the way to future research.

In order to answer the question *How is meaning derived from sound?,* the work researched a range of disciplines covering both the derivation of meaning from sound and the design of meaning in sound. The conclusions of this investigation are included in the model as a range of perceptual functions (such as affordances and modes of listening) and the effects of player experience and context in addition to the concepts of auditory icon design and sonification. A new mode of listening (navigational listening) has been defined to account for the navigational affordances of sound. In particular, the concept of sonification has led to a new way of conceptualizing the relationship between player and sound whereby the game engine is able to encode meaning into sound in order to communicate with the player and thereby provide a greater understanding of the gameplay than vision alone provides.

The work dealt with audiovisual FPS games and so the question was asked *What is the relationship between sound and image in the FPS game?* In answering this question, it has been demonstrated that concepts borrowed from Film Sound theory and other disciplines may be used or adapted for use within the FPS game acoustic ecology model. These include synchresis, acousmatic and visualized sound and the causality and indexicality of sound and, where necessary, these were adapted to account for the kinaesthetic nature of FPS games. It was shown that sound in the FPS game is strongly predicated upon player action, as initiated through haptic input, unlike sound in cinema which is predicated upon image. Furthermore, the concept of the first-person auditor was developed in order to position the player at the centre of the FPS game sound experience.

Because chapter 1 suggested that the player is a fundamental component of the acoustic ecology, the next question was *What is the role of sound in immersing the player in the acoustic ecology?* in order to provide a line of inquiry that might help explain how the player comes to be such a component. In pursuit of the answer, the work first investigated a range of possible sonic spaces in the acoustic ecology in which the player might be immersed and, in addition to utilizing pre-existing concepts of paraspace, developed the notion of *resonating space*. The player is physically immersed in a real resonating space and this is the only form of physical, sensory immersion which FPS games offer. Also defined were the terms *choraplast*, *topoplast*, *aionoplast* and *chronoplast* to explain the spatializing functions of sound within the game. To explain perceptual immersion in the FPS game acoustic ecology, the work first prepared the way by defining and examining a class of diegetic sounds. Consequently, the notions of *ideodiegetic* (comprising *exodiegetic* and *kinediegetic* sounds) and *telediegetic* sound were developed as an aid to differentiating sounds which the player hears immediately (classed as those triggered by the player and all other heard sounds) and those which the player cannot hear but to which other players react thus having later consequence for the first player. Existing theories and concepts of immersion in virtual environments and soundscape theory were applied to the acoustic ecology model. In doing so, the work has been able to state that FPS game diegetic sounds may be classed as keynote or signal sounds, may have, either singly or in combination, a variety of immersive functions and may be classified as attractors, connectors or retainers depending upon which immersive function is given priority by the player.

Finally, it was asked *What are the relationships between players on the basis of the sounds they trigger?* To answer this, the classifications of diegetic sound were used, in particular, telediegetic sound and the work has been able to show that there is a complex, relational web based on sound that exists between players in a multiplayer FPS game that extends not solely over geographical space but also, through the effects of telediegesis, acts over time. It has been demonstrated that some aspects of these relationships may be discussed and understood through diverse concepts when viewed through the lens of the acoustic ecology model and framework. In particular, the concepts of dramatic performance and autopoiesis were applied to the discussion of such relationships.

There are a number of weaknesses or things left unsaid which may be identified in the acoustic ecology model and these prove useful in identifying possible future research. Furthermore, the research presented in this work raises a host of interesting questions, speculations and avenues of future inquiry some of which are summarized below.

The model of the acoustic ecology is devised solely to account for the player experience of sound in FPS games. Throughout this work, though, it has been suggested that it might prove valuable to an understanding of sound in digital game genres other than FPS. This, then, is a further line of investigation that may be pursued. It may be that the framework and model can be expanded to incorporate new concepts or that they are modified to account for other digital game paradigms. For instance, *how do the spatializing functions of sound, if any, work in the context of a third player perspective game?* or *what are the sonic relationships, if any, between players in a MMORPG?*

Additionally, the work has deliberately mitigated the role of nondiegetic sound in the FPS game, in particular the role played by the musical soundtrack. Other researchers may argue that such music (in the FPS game) is diegetic in that it does have an affect upon the actions of players in the game. I myself have suggested, from personal experience, that this is indeed often the case but have been able to exclude music from any role in the acoustic ecology model firstly by stating that the work limits itself to a discussion of diegetic sound, secondly, by arguing that diegetic sound derives solely from objects and actions within the game world (and so does not include the musical soundtrack) and, thirdly, by asserting that the analysis of music requires a substantially different methodology than that required for the analysis of non-musical sound. An investigation of the role of the soundtrack in FPS games, the player's experience of it and an attempt to incorporate such findings into the model is, then, a task for future research.

This work has been theoretical in its approach. An important task is to support the hypothesis with empirical and experimental methods in order to vigorously test the conceptual framework and model, adapting them as necessary. Although the work is based around the player's understanding of FPS game sound, the player's voice, other than my own, is not heard. Empirical research, therefore, should include qualitative and quantitative data (gathered through questionnaires and observation, for example) aiming to divulge players' experiences of FPS game sound. Additionally, a comprehensive case study of an FPS game, building upon the illustrative examples I have provided, would be beneficial in pursuing the hypothesis further to the point, perhaps, of forming a fully-fledged theory of the FPS game acoustic ecology.

A range of FPS games (notably *Quake III Arena* (id Software, 1999)) were used to exemplify aspects of the conceptual framework and to broaden its potential application, but the final chapters were illustrated using *Urban Terror* (Silicon Ice, 2005). Although, in many respects, a typical FPS run and gun game, as a realism mod it differs in some aspects to more fantastical FPS games, such as *Quake III Arena* or others of the *Quake* series. It has already been noted the greater use of nomic auditory icons in *Urban Terror* as a point of difference to *Quake III Arena* suggesting that this mirrors the aspirations to realism of the former. However, other differences that, at first sight, have little to do with sound may, in fact, have significance for future elaborations of the framework and model. For example, in *Quake III Arena* the visual environment is more imaginary (that is, it is less a simulation of the architecture and locations to be found on Earth) than that of *Urban Terror* and there are certain actions and objects in the former that are not possible or that do not (yet) exist in reality — operation in space without pressure suits, teleporters and jump pads, for example. The question needs to be asked: how does one design a sound for an imaginary object or action in such a way that the player will not be presented with a completely impenetrable or absurd soundscape that militates against immersion?

Quake III Arena deals with the issue in many cases by using metaphor and caricature (symbolic and metaphorical auditory icons) or by using existing sonic conventions from science fiction cinema — for example, bouncing on the jump pad produces a sound that may be onomatopoeically described as 'boing'. This elastic sound is akin to the sound produced when bouncing on a trampoline or a pogo-stick in the real world. Still, though, the sound is a realistic one in that a real-world object or action can be visualized to match it (hence the visualization of the trampoline or the pogo-stick). What are the opportunities to provide fantastic sounds with fantastic imagary? Perhaps, in the more fantastic, imaginary worlds of non-realism FPS games, realist sound is required to anchor the game world in a simulacrum of reality such that the player may 'respond to it as if it were real'. Mis-matching sound to 'real', photographic image often produces absurd, comical effects (the films of Jaques Tati for example); juxtaposing a nomic sound to a cartoon image seems to produce not a laugh but, rather, serves to root that image in reality, to give it gravitas.[1] This is one line of questioning that may be followed and that may, investigating the more fantastical FPS game, lead to further insights to be added to the framework and model.

---

[1] I have already noted in chapter 7 that while computer graphics can produce stunningly realistic animations (witness the recent work of Weta Workshops), voice synthesis lags far behind. Two reasons might be suggested for this, both having to do with the generally anthropomorphic nature of cinematic animations — from Mickey Mouse to Gollum and King Kong, animators tend to strive to give human qualities to their character's features and movements. Firstly, and as pure conjecture, it may be that humans are more prepared to perceptually paper over visual cracks than sonic anomolies in the animation given the sensitivity of the human auditory system. Secondly, it may be that it is the voice and its expression of language (with all the conceptual and cognitive faculties implied) that is the primary marker of the human being as opposed to other species — recognition of this may be found in stories from Christianity to Mayan *Popul Vuh* creation myths, for example — using a real human voice dubbed onto an animation strengthens its anthropomorphic nature far more than any other factor. This may, in fact, work in reverse. Adding a too-near-to-perfect synthesized human voice to an animation may be unsettling for the film spectator who expects the creatures displayed to be human-like, but not human which such a voice would make it (film credits indicate that voices are not synthesized in modern films but are dubbed and therefore the animations are made palatable to the audiences). In the same way that Adorno and Eisler (1994) suggested that music was added to early cinema as a parallel to 'whistling in the dark', real human voices are (obviously) dubbed onto animations to dehumanize them and to maintain the splendid isolation of our species.

Furthermore, the work has frequently suggested interesting avenues of thought that, while not entirely germaine to the hypothesis, might be followed up in the future. Some of these include: *the need for simulation or emulation of reality in the soundscape* and, following on from the brief discussion in chapter 9, *the FPS game acoustic ecology as theatre.* The question about simulation or emulation is particularly interesting in the case of the realism FPS game. In addition to constraining actions to real-world scenarios, the (game) developmental emphasis seems to be on making the image as photographic as possible and packing in as many (simultaneous) nomic sounds as possible. Given the constraints of computer processing power when playing an FPS game, it may well be that, in order to perceptually immerse the player in the acoustic ecology (and, therefore, the game world), it is not necessary to provide an acoustic ecology that emulates real-world acoustic ecologies. A deliberately reduced sonic verisimilitude (perhaps using synthesis rather than audio samples), that carefully follows the model I have provided, may be all that is required.

Ultimately, though, following publications resulting from this research, it is to be hoped that other researchers will adapt and apply the conceptual framework and model formulated here to analyses of digital games, not necessarily solely in the area of sound,[2] and that this will validate the underlying impetus of the work: that sound in digital games is worthy of greater attention than it has attracted to date.

Coda

The image displayed on the screen in front of you is synchretically combined with the real resonating space that physically envelops you as you sit in front of the computer. Within this sonified, resonating space, the visualized sounds of a virtual resonating space confirm the apparent scale and depth of the spaces represented on screen and provide other affordances. There is a world beyond the one shown. A reverberating organ indicates the resonating space and locational paraspace of a church to your right and other environment sounds inform you that you are immersed within a rural paraspace. Experience dictates that, having just entered this world, the busy character and interactable sounds that you hear all around come from your teammates as they ready themselves for battle and this is backed up by radio signals providing feedback on your team leader's intentions. Utilizing navigational listening, you kinaesthetically follow the connecting audio beacon of the organ music to its source and linger for a while enjoying the retaining qualities of the Bach fugue which reduced and semantic listening modes afford. Soon however, distant acousmatic sounds require a switch to causal listening and, discerning the sounds of a firefight, you turn to navigational listening once more and move to join the fray. The kinediegetic, exteroceptive sounds of your footsteps are quickly overshadowed by a kinediegetic, proprioceptive panting forcing you to slow to a walk as the exodiegetic sounds of battle intensify and the (by now acousmatic) organ music is kinaesthetically attenuated. While this has been occurring, in another part of the game world, a team member has darted into the enemy's base and snatched their flag and this unseen event has been signalled by an alarm. Enemy soldiers have been pursuing your plucky teammate, following the sounds of his footsteps and gunfire as he desperately weaves in and out of the abbey's cloisters. You are met by

2 Perhaps the ecology of the (FPS) digital game?

the telediegetic consequences of this when the dramatic chase appears on the screen in front of you in the form of your bleeding teammate and the frantic enemy. A few clicks of haptic input send a couple of grenades the enemy's way and almost immediately the signal sounds of explosions and cries of the wounded pleasingly mingle with the more bucolic keynote sounds of the soundscape. An enemy experiences first-hand the effects of your telediegetic advance to battle as your shotgun explodes in her face allowing you to respond to the flag-carrier's radioed requests for a medic whilst ignoring the blood gushing from your own wound. Accompanying the flag-carrier to your team's base, the flag is captured with a global fanfare (much to the distress of the enemy) as radioed feedback from your dispersed team members congratulates you on your heroics.

Appendix 1

The FPS Game *Urban Terror*

A. The game

Urban Terror (Silicon Ice, 2005) is a total conversion modification (mod) for *Quake III Arena* (id Software, 1999). The first beta version of the game was released in August 2000 and, although at time of writing the development of the mod continues, the current version is 3.7 released in November 2004. As a mod of *Quake III Arena*, it must comply with id Software's *End User License Agreement* (based on the *GNU General Public License* (Free Software Foundation, 1989)) which, among other specifications, stipulates that without the payment of a license fee, mods may modify parts of the *Quake III Arena* game engine code as long as the result is made freely available to the public.

There are many other non-commercial mods of *Quake III Arena* which bear a great deal of similarity to the original game, for example *Rocket Arena* (2000-2006). However, as a total conversion mod, *Urban Terror*, visually, aurally and in many aspects of gameplay, bears little resemblence to *Quake III Arena*, so much so that a player unfamiliar with *Urban Terror* would not recognize that its underlying technology is *Quake III Arena*. Perhaps the most striking difference is found in the premise of the two games which in turn strongly affects the looks, feel and sounds of each game. *Quake III Arena*, although never explicitly stated, is set in an otherworld where the architecture is a mix of Gothic and technological metallic, where there is a range of characters from fantastical to human, items not found in reality, such as teleporters and quad damage, and weaponry which spans real-world weapons through to arms which would not be out of place in a futuristic science fiction film. *Urban Terror*, by contrast, is set on (a recognizable) Earth (in fact some of its levels are designed following geographical locations such as Siberia or more general locations such as an abbey), has only humans for its character models and uses only weaponry which has been scrupulously modeled, both visually and aurally, after their real-world equivalents.

It is these features which have led to *Urban Terror* sometimes being described as a realism mod (Ambrosetti, n.d., for example). However, despite some efforts at a simulation of various aspects of reality, it should be borne in mind that there are many features (such as the characters' tolerance for weapons fire and their ability to quickly heal themselves) which remain unreal or which have no parallel in reality. As PainBerry (n.d) writes describing the developers' approach to designing the game:

> The team definately [*sic*] seem to be picking up and responding to the shortcomings of realism games to date, whether they know it or not. Even their weapons specialist (who you would think would be obsessed with instigating realism redundancies) has figured out that random offset on weapons totally undermines skill. Many of their mappers understand how much more important gameflow is than eyecandy (althought [*sic*] eyecandy is great and all). They're taking a good

> perpective [*sic*] with their game design - if it's realistic, and it benefits the game, it's in, otherwise they sidestep it. They kept in strafe jumping ferchrissake! This is realism grown to maturity, with the idio(t)syncracies of their parents breeded [*sic*] out.

This theme of a tempered version of reality is backed up by the designers themselves. As WU states: "Urban Terror is a total conversion that aims to achieve the realistic sense of suspense and action that real life situations would provide. Although its primary inspiration is realism, we are not forgetting the community we are developing for" (quoted in Law, n.d.). And this is hyperbolically restated by GottaBeKD:

> "Urbanites" will get lost in the reality of it all. From the beautiful present day locations to the frighteningly real models, there will be tons of things working together to give our players that action fix that they all need (quoted in Law, n.d.).

There is a sense, then, that the FPS game is no game if it is too real: "So, is realism possible in any Quake game? Well, now you're just talking simulations. Is a realism/game cross over [*sic*] possible in any Quake game? With a little decency, common sense and taste, hell yeah!" (PainBerry, n.d.).

Urban Terror displays all the key characteristics of the modern FPS game. Although third-person perspective is supplied when spectating other players (that is, while not playing the game), the mode of play is first-person perspective, as indicated by a pair of hands holding a weapon and receding perspectivally away from the player into the screen. Within the game environment, the character has almost complete freedom of movement within the bounds set by the level's architecture and there is a simulation of gravity (including character damage if jumping from too great a height). Visually, the game environment makes use of FPS game techniques and characteristics such as parallax, depth scaling and perspective and this is matched aurally with multichannel localization and depth cues all contributing to a set of resonating spaces (both real and virtual) and paraspaces which themselves contribute to the illusory 3-dimensionality of the FPS game world in which the player is not only the first-person viewer but is also the first-person auditor.

Like almost all FPS games, one of the main aims of *Urban Terror* is to kill game opponents and to avoid damage to or the death of the player's character (the relative priority of this in relation to other aims is detailed below). The opponents may be either bots in single-player configuration or they may be other players' characters in network-based (LAN or Internet) multiplayer configurations. Players are limited in the weaponry and items they may carry with them unlike *Quake III Arena* in which the player may collect and use all available weapons. There are fourteen weapons, one of which is a knife and two of which are pistols (termed sidearms), covering a range of the small arms firepower found in modern warfare (for example a grenade launcher, sniper rifles, shotgun and machineguns of different calibres and rates of fire).

Additionally, there is a range of items including hand grenades (explosive or smoke), laser sights, silencers, medkits, night-vision goggles and kevlar armour that may be carried and used by the character. For some items their use is automatic, for others, their use can be toggled on and off. A character must always carry a knife and sidearm and may choose up to four other objects but only a maximum of two weapons other than the knife and sidearm. The types of items a character carries and uses has an effect not only upon their role within the game (especially within team games) but also has an effect upon their playing style in the game. For example, the medkit allows that player to heal other team members (and themselves) faster and to a greater degree than when healing without the medkit; the kevlar armour and helmet may provide greater protection but mean that the character may not run as fast as others who are not weighed down by the armour.

In addition to the typical FPS game modes based on run and gun scenarios, such as deathmatch (or free for all) and team deathmatch, *Urban Terror* offers five more team-based game modes (some of which, but not all, are found in other FPS games as well such as *Quake III Arena*, *Return to Castle Wolfenstein* (Gray Matter Studios & id Software, 2001) and *Counter-strike* (Valve Software, 1999)). These are:

- *Team Survivor.* This is a last-man-standing game mode and characters that die do not respawn until the next round.
- *Follow the Leader.* The team's leader must be protected and the leader can score points by touching the enemy's flag. Characters that die do not respawn until the next round.
- *Capture and Hold.* A number of bases may be occupied and held by team members. Teams score points at minute intervals for each base they hold.
- *Capture the Flag.* Teams must capture the enemy's flag and return it to their own base (which must still contain their team's flag) in order to score points.
- *Bomb.* A team plants a bomb in one of several target areas and defends that area whilst the other team either attempts to stop the bomb being planted or attempts to defuse the planted bomb within a specified time limit.

In much of its scenario then, *Urban Terror* presents objects and actions which exist or may be undertaken in the real world although realism in the game is not pursued to the detriment of gameplay. The game is a typical example of the modern FPS run and gun game (as exemplified by its first-person perspective, its 'hunter and the hunted' premise and its game modes, for example) and one, therefore, which is appropriate to use as an illustrative example when discussing the conceptual framework and the FPS game acoustic ecology.

B. Audio samples

During the process of coding a mod for *Quake III Arena*, developers may choose to use the original audio samples, to use a mix of original audio samples and new audio samples or to use solely new audio samples. Because the premise of *Urban Terror* is so different to *Quake III Arena* — a realism mod as opposed to a fantastical game world — the developers have chosen the latter option of creating a total of over 850 new audio samples to form the basis of the acoustic ecology of *Urban Terror.* This section briefly summarizes these audio samples, discusses the requirements and capabilities of the *Quake III Arena* game audio engine and analyzes the audio samples from the point of view of their organization within the game code.

In *Urban Terror* there are 853 base audio samples[1] stored in the directories sound/ and music/ (there is a directory Sounds/ with a subdirectory containing eight audio samples but, according to the sound designer, the inclusion of this was an error when packaging the game (Klem, personal communication, 2004)). Of these 853 audio samples, there is one file in music/, which is the title music used for the menu interface, and five audio samples in sound/misc/ indicating menu operations such as clicking and selecting a menu option; all other audio samples are available to be used during gameplay. These remaining 847 audio samples are organized into 44 directories with nine audio samples being at the top level of the sound/ directory. At least 19 of these 44 directories are specific to *Urban Terror* core levels (levels that come with the game package as opposed to levels designed by users)[2]. There are a further 13 directories providing a range of audio samples that are available to be used in various levels (core levels or user levels). The 12 remaining directories contain the base audio samples that are directly used by the game engine in response to player-object interactions and other actions or game states (such as game status messages). These are termed game-specific audio samples as opposed to level-specific audio samples.

This storage organization is one that reflects the somewhat haphazard nature of modding with multiple developers and over several version releases of the game; updating *Urban Terror* is a matter of downloading incremental updates to be added to the existing installation such that no files are ever removed regardless of whether they continue to be required in the later version or not. There is duplication of some audio samples, audio samples and directories belonging to levels that are not in later versions of the game, audio samples added by mistake (the Sounds/ directory for example) or audio samples simply there for humorous purposes but not intended to be used during gameplay; discussing the audio sample sound/urban-terror/tehdrunkenpilot.wav, Klem states: "This file was actually put in there on purpose as an easter-egg which may only be found by poking through the pk3". Additionally, the *Quake III Arena* game audio engine only supports WAV files sampled with a maximum sampling frequency of 22.05kHz[3] yet some audio samples are sampled at 44.1kHz; again, according to Klem, this was an error.

These 12 directories contain 598 audio samples; including the nine audio samples residing at the top level of the sound/ directory, this produces a total of 607 game-specific audio samples. As a reminder, some of these are superfluous audio samples for a variety of reasons enumerated above. Some of the 12 directories have subdirectories, for example the sound/player/ and sound/radio/ directories each have subdirectories for male and female audio samples (the sound/player/ directory also has a footsteps/ directory) whilst the weapons/ directory contains a subdirectory for each of the available weapons and some of the items (such as the two types of grenades).

---

1 As opposed to user-created audio samples which are distributed with user-created levels.

2 Some of these directories are remnants of levels available in older versions of the game but no longer available in the latest version.

3 Compact discs are sampled at 44.1kHz.

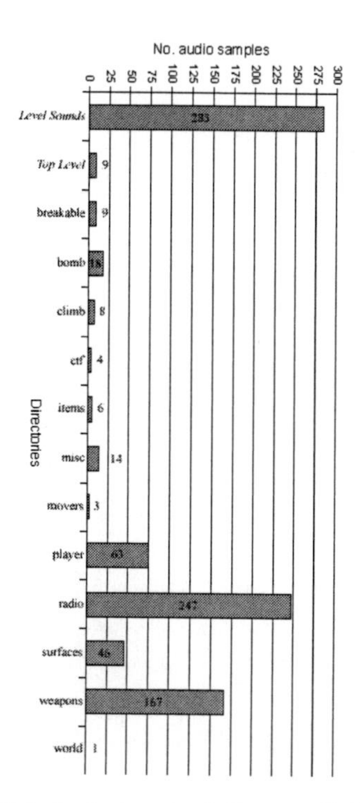

Figure A1.1. Audio samples in the sound/ directory.

Figure A1.1 demonstrates the distribution of audio samples within the sound/ directory (for clarity, all level-specific audio samples have been collated into one item). Of interest here is the disparity in distribution of audio samples across the game-specific sound/ directories. The sound/radio/ directory, for example, has 247 audio samples within it whilst the sound/weapons/ directory has 167 sounds leaving the remaining 293 game-specific audio samples to be shared among 10 directories.

Appendix 2

Glossary

A glossary of the major terminology and neologisms used throughout this work. The page numbers indicate the point in the work where the term first appears or is first explained or defined.

acousmatic sound A sound whose source cannot be seen. In the FPS game, such sounds indicate the existence of parts of the game world or the actions of players or game objects beyond those that can be seen on the screen. **75**

affordance In terms of sound, the opportunity (or opportunities) offered by it. The prioritizing of affordances is subject to player experience. **55**

aionoplast A sound whose function is to contribute to the perception of a temporal paraspace in the FPS game. **109**

ambient sound An environment sound played by the game engine and which is often used to indicate resonating space or locational paraspace. It may be kinaesthetically interactable or globally pervasive. **57**

attractor A sound whose function is to tempt the player into some action. **128**

audiation Mapping an imagined sound to a mute image. **62**

audification See **sonification**. **62**

audio sample Audio data that is a digital representation of a soundwave and therefore a method of storing a representation of a recorded sound. Almost all modern FPS games use audio samples rather than synthesized sound. **5**

bot A computer-generated character in an FPS game. **9**

causal listening mode A mode of listening whose focus is on obtaining information about the cause of the sound. **52**

choraplast A sound whose function is to contribute to the perception of a resonating space in the FPS game. **103**

chronoplast A sound whose function is to indicate the passage of chronological time. **109**

connector A sound used by the player for orientation in the game world which is normally attended to with the **navigational listening mode**. **128**

cross-modality The relationship between different sensory and perceptual systems particularly, in the case of this work, sight and hearing. **70**

diegetic sound As adapted from film sound theory for the FPS game acoustic ecology, sound that originates in the game world from its environment, objects and characters and which is affected and effected by that environment and the objects and characters. **116**

digital signal processing (DSP) Manipulation and processing of digital data. In the context of audio samples, such processing may include the artificial addition of reverberation, equalization or amplification of the sample. **8**

exodiegetic sound An **ideodiegetic** sound that is not **kinediegetic**. Such sounds, as heard by one player, are the sounds of other players or **bots**, environment sounds or other game object sounds that have not been triggered directly by that player. **119**

external auditor In cinema, an imaginary person listening to the action on-screen but who is external to that story world. This is achieved by the placement of a microphone in the same position as the camera and was an attempt to match sound scale to image scale. **81**

first-person auditor A term devised to explain the auditory status of the FPS game player by analogy with similar terms as used in cinema. The camera and microphone are co-located but, importantly, the first-person auditor is not only within the world displayed on the screen but exercises kinaesthetic control over much of that world. **83**

iconic sound A sound having no direct, causal relationship to the object or action with which it is associated. In reality, the object or action could not have caused the sound. **87**

ideodiegetic sound A **diegetic** sound which the FPS player can hear. Ideodiegetic sounds are either **kinediegetic** or **exodiegetic**. **118**

impossible auditor In cinema, this is an imaginary character, who, unlike humans, is able to hear all events portrayed on the screen with similar intensity and clarity. This was achieved by the positioning of multiple microphones in the film set but, in the early days of talkies, was bedeviled by rudimentary sound balancing techniques. **81**

indexical sound A sound having a direct, causal relationship to the object or action with which it is associated. There is a direct relationship between sign and signifier. **87**

internal auditor In cinema, an imaginary listener positioned within the world of the events portrayed on-screen. A microphone is positioned close to the actors, away from the camera, in an attempt to improve the intelligibility of dialogue at the expense of a correct correlation between sound intensity and image size. **81**

isomorphic sound Typically in film, a sound or piece of music which has an analogous relationship to the rhythm and/or movement seen on-screen. Also known as *mickey-mousing*. **87**

keynote sound From soundscape theory. A sound forming part of the sonic background of the soundscape and therefore not necessarily consciously attended to. See **signal sound**. **110**

kinaesthetics The ability of a player in the FPS game to turn (his character) towards the source of an **acousmatic** sound and one of the distinguishing features of FPS games contrasted to cinema. **25**

kinediegetic sound An **ideodiegetic** sound triggered directly by the player's action in the FPS game. **119**

navigational listening mode A mode of listening whose focus is on the use of sound as audio beacon, guiding the FPS player to its source. **32**

network auralizer A **sonification** system in which sound is used to monitor the status of a computer network. **64**

nomic auditory icon A sound having a strongly causal relationship to the object or action it represents. **67**

quasi-causal sound A sound having a degree of causality which hints at its source object or action enabling the listener to divine that source with experience at which point the sound may be said to be causal. **77**

paraspace An acoustic space indicating location or temporal period in the FPS game. **93**

perceptual realism A constructed reality, typically in virtual environments, that is persuasive enough in its representation to allow the user or player to respond to it as if it were real. **120**

reduced listening mode An appreciation of the qualities of sound without reference to cause or meaning. **52**

resonating space An acoustic space having volume and related to the dimensions and materials of spaces in the FPS game. The FPS game has at least two resonating spaces, one real (that envelops the player) and at least one virtual (that represents universal space(s) in the game world). **93**

retainer A sound whose function is to cause the player to linger in a part of the game world. **128**

schizophonia A physical separation or dissociation of sound from sound source. Typically a modern phenomenon which is enabled by recording and telephony. **84**

semantic listening mode The use of a semiotic code to interpret the meaning of sound. **52**

signal sound From soundscape theory. A foreground sound in the soundscape that is consciously attended to. See **keynote sound**. **110**

sonification The translation of audio or non-audio data into sound. 0th order sonification (or **audification**) is a direct translation of audio data (for example, an **audio sample**) into sound whereas other forms of higher-order sonification translate parameters of non-audio data into parameters of sound. For example, population density may be mapped to sound intensity. **62**

sonography When assessing the FPS game (acoustic ecology) as dramatic performance, the sonography is the aural equivalent of scenography. **159**

soundmark A sound which is an identifying feature of a locale. An analogy with landmark. **110**

symbolic auditory icon A sound having an arbitrary relationship to the object or action it represents. **59**

telediegetic sound A sound heard by and responded to by one player, the actions of that response having later consequence for another player. An important instance of sonic relationships in a multiplayer game. **118**

topoplast A sound whose function is to contribute to the perception of a locational paraspace in the FPS game. **105**

visualization A pictorial imagining of an object or an action for a sound. Also, the de-acousmatization of sound. **79**

Bibliography

Aarseth, E. (2001). Computer game studies, year one. *Game Studies, 1*(1). Retrieved August 31, 2006, from http://www.gamestudies.org/0101/editorial.html.

Aarseth, E., Smedstad, S. M., & Sunnanå, L. (2003, November 4—6). *A multi-dimensional typology of games*. Paper presented at Level Up, Utrecht Universiteit.

Adorno, T. W., & Eisler, H. (1994). *Composing for the films*. London: Athlone Press.

Affordance. (2004-2006). *Wikipedia*. Retrieved March 20, 2006, from http://en.wikipedia.org/wiki/Affordance.

Aigaion. (2004-2006). (Version 1.1) [PHP].

Altman, R. (1992c). General introduction: Cinema as event. In R. Altman (Ed.), *Sound Theory Sound Practice* (pp. 1—14). New York: Routledge.

Altman, R. (1992b). Sound space. In R. Altman (Ed.), *Sound Theory Sound Practice* (pp. 46—64). New York: Routledge.

Altman, R. (Ed.). (1992a). *Sound theory sound practice*. New York: Routledge.

Ambroise, C. (2005-2006). dokuwikibibtexplugin. [Dokuwiki plug-in].

Ambrosetti, F. (n.d.). Urban Terror 2.6 released. Retrieved June 2, 2006, from http://www.gamershell.com/news/4383.html.

Anderson, J. D. (1996). *The reality of illusion: An ecological approach to cognitive film theory*. Carbondale and Edwardsville: Southern Illinois University Press.

Atari. (1972). Pong. [Computer Game]. New York: Atari.

Back, M., & Des, D. (1996). Micro-narratives in sound design: Context and caricature in waveform manipulation. Retrieved March 12, 2004, from http://www2.hku.nl/~audiogam/ag/articles/micronaratives.htm.

Ballas, J. A. (1994). Delivery of information through sound. In G. Kramer (Ed.), *Auditory Display: Sonification, Audification, and Auditory Interfaces* (pp. 79—94). Reading MA: Addison-Wesley.

Barthes, R. (1967). The death of the author. *Aspen, 5+6*. Retrieved December 18, 2006, from http://www.ubu.com/aspen/aspen5and6/threeEssays.html#barthes.

Bartkowiak, A. (Director) (2005). *Doom III* [Film]. J. Wells (Producer). USA: Universal Pictures.

Böhme, G. (2000). Acoustic atmospheres: A contribution to the study of ecological acoustics. *Soundscape, 1*(1), 14—18.

Benjamin, W. (1936). The work of art in the age of mechanical reproduction. Retrieved March 8, 2006, from http://www.marxists.org/reference/subject/philosophy/works/ge/benjamin.htm.

Bernds, E. (1999). *Mr. Bernds goes to Hollywood: My early life and career in sound recording at Columbia with Frank Capra and others*. Lanham MD: Scarecrow Press.

Bernstein, D. (1997). Creating an interactive audio environment. *Gamasutra*. Retrieved September 6, 2004, from http://www.gamasutra.com/features/19971114/bernstein_01.htm.

Berry, C. (1987). *The actor and his text*. London: Harrap.

Bioware. (2002). Neverwinter Nights. (Version Premium Modules) [Computer Game]. Lyon: Infogrames.

Bird, B. (Director) (2004). *The Incredibles* [Film]. J. Walker (Producer). USA.

Blattner, M. M., Sumikawa, D. A., & Greenberg, R. M. (1989). Earcons and icons: Their structure and common design principles. *Human-computer Interaction, 4*, 11—44.
Boulanger, C. (2006). Bibliograph. (Version 0.9) [Python/PHP/Javascript].
Boyd, A. (2003). When worlds collide: Sound and music in films and games. *Gamasutra*. Retrieved September 6, 2004, from http://www.gamasutra.com/features/20030204/boyd_01.shtml.
Brandon, A. (2005). *Audio for games: Planning, process, and production*. Berkeley: New Riders.
Breinbjerg, M. (2005). The aesthetic experience of sound: Staging of auditory spaces in 3D computer games. Retrieved January 24, 2006, from http://www.aestheticsofplay.org/breinbjerg.php.
Bridgett, R. (2003b). Audio pre-design. Retrieved March 8, 2004, from http://www3.sympatico.ca/qualish/pre.htm.
Bridgett, R. (2003a). Off screen sound in interactive media. Retrieved September 9, 2003, from http://www3.sympatico.ca/qualish/off.htm.
Bross, M. (1997a). Effective music for games and multimedia. *Gamasutra, 1*. Retrieved September 6, 2004, from http://www.gamasutra.com/features/sound_and_music/111497/effective_musi c.htm.
Bross, M. (1997b). Powerful audio tips for game and multimedia developers. *Gamasutra*. Retrieved September 6, 2004, from http://www.gamasutra.com/features/sound_and_music./103197/audio_tips.ht m.
Bruner, J. S. (1957). On perceptual readiness. *Psychological Review, 64*(2), 123—152.
Bungie Studios. (2001-2004). Halo series. [Computer Game]. Redmond: Microsoft.
Burn, A., & Parker, D. (2003). *Analysing media texts*. London: Continuum.
Burn, A., & Schott, G. (2004). Heavy hero or digital dummy? Multimodal player-avatar relations in *Final Fantasy 7. visual communication, 3*(2), 213—233.
Bussemakers, M. P., & de Haan, A. (1998, November 1—4). *Using earcons and icons in categorisation tasks to improve multimedia interfaces*. Paper presented at 5th International Conference on Auditory Display, Glasgow.
Cameron, J. (Director) (1991). *Terminator 2: Judgement day* [Film]. J. Cameron (Producer). USA: Tri-Star Pictures.
Carr, D. (2006b). Games and narrative. In *Computer Games: Text, Narrative and Play* (pp. 30—44). Cambridge: Polity.
Carr, D. (2006a). Space, navigation and affect. In *Computer Games: Text, Narrative and Play* (pp. 59—71). Cambridge: Polity.
Cavalcanti, A. (1939). Sound in films. Retrieved January 16, 2006, from http://lavender.fortunecity.com/hawkslane/575/sound-in-films.htm.
Chagas, P. C. (2005). Polyphony and embodiment: A critical approach to the theory of autopoiesis. *Revista Transcultural de Música, 9*. Retrieved July 7, 2006, from http://www.sibetrans.com/trans/trans9/chagas.htm.
Chion, M. (1994). *Audio-vision: Sound on screen* (C. Gorbman, Trans.) New York: Columbia University Press.
Collins, K. E. (n.d.). Video games audio. Retrieved November 19, 2003, from http://www.dullien-inc.com/collins/texts/vgaudio.pdf.
Copier, M. (2003, November 4—6). *The other game researcher: Participating in and watching the construction of boundaries in game studies*. Paper presented at Level Up, Utrecht Universiteit.

Coppola, F. F. (Director) (1972). *The Godfather* [Film]. F. F. Coppola (Producer). USA: Paramount Pictures.

Corner, J. (1992). Presumption as theory: 'realism' in television studies. *Screen, 33*(1), 97—102.

Coward, S. W., & Stevens, C. J. (2004). Extracting meaning from sound: Nomic mappings, everyday listening, and perceiving object size from frequency. *The Psychological Record, 54*(3), 349—364.

Curtiss, S. (1992). The sound of early Warner Bros. cartoons. In R. Altman (Ed.), *Sound Theory Sound Practice* (pp. 191—203). New York: Routledge.

Cyan. (1993). Myst. [Computer game]. Brøderbund.

Czyzewski, A., Rostek, B., Odya, P., & Zielinski, S. (2001). Determining influence of visual cues on the perception of surround sound using soft computing. *Lecture Notes in Artificial Intelligence, 2005*, 545—552.

Darley, A. (2000). *Visual digital culture: Surface play and spectacle in new media genres.* London: Routledge.

Dhomont, F. (2004). Acousmatic update. *Sonic Arts Network.* Retrieved February 16, 2006, from
http://www.sonicartsnetwork.org/ARTICLES/ARTICLE1996DHOMONT.html.

Diegesis. (2003-2006). *Wikipedia.* Retrieved January 12, 2006, from http://en.wikipedia.org/wiki/Diegetic.

Digital Illusions. (2002). Battlefield 1942. [Computer Game]. Electronic Arts.

Doane, M. A. (1980). Ideology and the practice of sound editing and mixing. In T. de Lauretis & S. Heath (Eds.), *The Cinematic Apparatus* (pp. 47—56). London: Macmillan.

Dodge, C., & Jerse, T. A. (1985). *Computer music: Synthesis, composition, and performance.* New York: Schirmer Books.

Epic Games, & Digital Extremes. (2004). Unreal tournament 2004. [Computer Game]. New York: Atari.

Ermi, L., & Mäyrä, F. (2005, June 16—16). *Fundamental components of the gameplay experience: Analysing immersion.* Paper presented at Changing Views -- Worlds in Play, Toronto.

Eskelinen, M. (2001). The gaming situation. *Game Studies, 1*(1). Retrieved August 31, 2006, from http://www.gamestudies.org/0101/eskelinen/.

Everest, F. A. (1984). *Acoustic techniques for home and studio* (2nd ed.) Blue Ridge Summit, PA: Tab Books.

Familant, M. E., & Detweiler, M. C. (1993). Iconic reference: Evolving perspectives and an organizing framework. *International Journal of Man-Machine Studies, 39*(5), 705—728.

Fan fiction inspired by Halo. (n.d.). Retrieved April 13, 2006, from http://halosn.bungie.org/fanfic/.

Fencott, C. (1999). Presence and the content of virtual environments. Retrieved August 4, 2005, from
http://web.onyxnet.co.uk/Fencott-onyxnet.co.uk/pres99/pres99.htm.

Fernández-Vara, C., Zagal, J. P., & Mateas, M. (2005, June 16—20). *Evolution of spatial configurations in videogames.* Paper presented at Changing Views -- Worlds in Play, Toronto.

First-person shooter. (2002-2006). *Wikipedia.* Retrieved April 12, 2006, from http://en.wikipedia.org/wiki/First-person_shooter.

Fitch, W. T., & Kramer, G. (1994). Sonifying the body electric: Superiority of an auditory over a visual display in a complex, multivariate system. In G. Kramer (Ed.), *Auditory Display: Sonification, Audification, and Auditory Interfaces*

(pp. 307—325). Reading MA: Addison-Wesley.
Folmann, T. B. (n.d.). Dimensions of game audio. Retrieved November 23, 2004, from http://www.itu.dk/people/folmann/2004/11/dimensions-of-game-audio.html.
Free Software Foundation. (1989). GNU general public license. Retrieved June 8, 2006, from http://www.fsf.org/licensing/licenses/gpl.html.
Friberg, J., & Gärdenfors, D. (2004, June 3—5). *Audio games: New perspectives on game audio.* Paper presented at Advances in Computer Entertainment Technology '04, Singapore.
Gamer's Edge. (1992). Catacomb 3D. [Computer Game]. Shreveport LA: Softdisk Publishing.
Gardey, G. (2005). BibORB. (Version 1.3.3) [PHP].
Gaver, W. W. (1986). Auditory icons: Using sound in computer interfaces. *Human-computer Interaction, 2*, 167—177.
Gaver, W. W. (1993a). How do we hear in the world? Explorations in ecological acoustics. *Ecological Psychology, 5*(4), 285—313.
Gaver, W. W. (1993b). What in the world do we hear? An ecological approach to auditory perception. *Ecological Psychology, 5*(1), 1—29.
Georgantzas, N. C. (n.d.). Self-organization dynamics. Retrieved November 21, 2006, from http://www.systemdynamics.org/conferences/2001/papers/Georgantzas_2.pdf.
Gibson, J. J. (1966). *The senses considered as perceptual systems.* Boston: Houghton Mifflin.
Gomery, D. (1980). Towards an economic history of cinema: The coming of sound to Hollywood. In T. de Lauretis & S. Heath (Eds.), *The Cinema Apparatus* (pp. 38—46). London: Macmillan.
Gray Matter Studios, & id Software. (2001). Return to Castle Wolfenstein. [Computer Game]. Activision.
Gröhn, M., Lokki, T., Savioja, L., & Takala, T. (2001, January 22—23). *Some aspects of role of audio in immersive visualization.* Paper presented at Visual Data Exploration and Analysis VIII, San Jose.
Grimshaw, M. (2001a). Grim Shores 2: Devil's Island. [Computer Game].
Grimshaw, M. (2001b). Grim Shores 3: Atlantis. [Computer Game].
Grimshaw, M. (2003-2006). WIKINDX. (Version 3.4) [PHP].
Grodal, T. (2003). Stories for eye, ear, and muscles: Video games, media, and embodied experiences. In M. J. P. Wolf & B. Perron (Eds.), *The Video Game Theory Reader* (pp. 129—155). New York: Routledge.
Guy, W. (2003). Design with music in mind: A guide to adaptive audio for game designers. *Gamasutra.* Retrieved September 19, 2004, from http://www.gamasutra.com/resource_guide/20030528/whitmore_pfv.htm.
Hahn, J. K., Fouad, H., Gritz, L., & Lee, J. W. (1998). Integrating sounds and motions in virtual environments. *Presence, 7*(1), 67—77.
Halpern, D. L., Blake, R., & Hillenbrand, J. (1986). Psychoacoustics of a chilling sound. *Percept Psychophys, 39*, 77—77.
Hess, D. (Director) (2005). *Stranger* [Film]. USA.
Hitman series. (2000-2006). [Computer Game]. Eidos.
Howard, D. M., & Angus, J. (1996). *Acoustics and psychoacoustics.* Oxford: Focal Press.
id Software. (1993). Doom. [Computer Game]. Activision.
id Software. (1999). Quake III Arena. [Computer Game]. Activision.
id Software. (2004). Doom 3. [Computer Game]. Activision.

Innocent, T. (2003, May 19—23). *Exploring the nature of electronic space through semiotic morphism.* Paper presented at 5th International Digital Arts and Culture Conference, RMIT, Melbourne.
Jackson, P. (Director) (2001-2003). *The Lord of the rings trilogy* [Film]. New Zealand/USA: New Line Cinema.
Jackson, P. (Director) (2005). *King Kong* [Film]. J. Blenkin, C. Cunningham, P. Jackson & F. Walsh (Producers). New Zealand: Universal Pictures.
Johnson, S. (2001). *Emergence: The connected lives of ants, brains, cities and software.* London: Penguin Books.
Jones, M. R., & Yee, W. (1993). Attending to auditory events: The role of temporal organization. In S. McAdams & E. Bigand (Eds.), *Thinking in Sound: The Cognitive Psychology of Human Audition* (pp. 69—112). Oxford: Clarendon Press.
Juul, J. (2001). Games telling stories? A brief note on games and narratives. *Game Studies, 1*(1). Retrieved September 16, 2003, from http://www.gamestudies.org/0101/juul-gts/.
Kattenbelt, C., & Raessens, J. (2003, November 4—6). *Computer games and the complexity of experience.* Paper presented at Level Up, Utrecht Universiteit.
Kücklich, J. (2003). Perspectives of computer game philology. *Game Studies, 3*(1). Retrieved September 16, 2003, from http://www.gamestudies.org/0301/kucklich/.
King, G., & Krzywinska, T. (2002). Cinema/videogames/interfaces. In G. King & T. Krzywinska (Eds.), *Screenplay: Cinema/Videogames/Interfaces* (pp. 1—32). London: Wallflower Press.
Kramer, G. (1994). Some organizing principles for representing data with sound. In G. Kramer (Ed.), *Auditory display: Sonification, audification, and auditory interfaces* (pp. 185—221). Reading MA: Addison-Wesley.
Kramer, G., Walker, B., Bonebright, T., Cook, P., Flowers, J., & Miner, N., et al. (n.d.). Sonification report: Status of the field and research agenda. Retrieved September 1, 2005, from http://www.icad.org/websiteV2.0/References/nsf.html.
Kress, G., & van Leeuwen, T. (2001). *Multimodal discourse: The modes and media of contemporary communication.* London: Hodder Arnold.
Kress, G. (2003). *Literacy in the new media age.* London: Routledge.
Kristeva, J. (1984). *Revolution in poetic language* (M. Waller, Trans.) New York: Columbia University Press.
Larsson, P., Västfjäll, D., & Kleiner, M. (2001, October 4—5). *Do we really live in a silent world? The (mis)use of audio in virtual environments.* Paper presented at AVR II and CONVR 2001, Gothenburg, Sweden.
Lasseter, J. (Director) (1995). *Toy story* [Film]. B. Arnold (Producer). USA: Buena Vista.
Lastra, J. (1992). Reading, writing, and representing sound. In R. Altman (Ed.), *Sound Theory Sound Practice* (pp. 65—86). New York: Routledge.
Lastra, J. (2000). *Sound technology and the American cinema: Perception, representation, modernity.* New York: Columbia University Press.
Laurel, B. (1993). *Computers as theatre.* New York: Addison-Wesley.
Law, C. (n.d.). Urban Terror! *GameSpy.* Retrieved June 2, 2006, from http://archive.gamespy.com/legacy/spotlights/urbanterror_a.shtm.
Law, J. (1992). Notes on the theory of the actor network: Ordering, strategy and heterogeneity. Retrieved July 14, 2006, from http://www.lancs.ac.uk/fss/sociology/papers/law-notes-on-ant.pdf.

van Leeuwen, T. (1999). *Speech, music, sound.* London: MacMillan Press.

Liang, W., & Tan, M. C. C. (2002). Vision and virtuality: The construction of narrative space in film and computer games. In G. King & T. Krzywinska (Eds.), *Screenplay: Cinema/Videogames/Interfaces* (pp. 98—109). London: Wallflower Press.

LoBrutto, V. (1994). *Sound-on-film: Interviews with creators of film sound.* Westport, CT: Praeger.

Malaby, T. M. (2006). Stopping play: A new approach to games. *Social Science Research Network.* Retrieved August 11, 2006, from http://ssrn.com/abstract=922456.

Manninen, T. (2003). Interaction forms and communicative actions in multiplayer games. *Game Studies, 3*(1). Retrieved September 16, 2004, from http://www.gamestudies.org/0301/manninen/.

Martin, R. L., Thrift, N. J., & Bennett, R. J. (Eds.). (1978). *Towards the dynamic analysis of spatial systems.* London: Pion.

Maturana, H., & Varela, F. (1980). *Autopoiesis and cognition: The realization of the living.* Dordecht: D. Reidel Publishing Co.

Maturana, H., & Varela, F. (1992). *The tree of knowledge: The biological roots of human understanding* (R. Paolucci, Trans.) Boston: Shambhala.

McAdams, S. (1993). Recognition of sound sources and events. In S. McAdams & E. Bigand (Eds.), *Thinking in Sound: The Cognitive Psychology of Human Audition* (pp. 146—198). Oxford: Clarendon Press.

McCloud, S. (1993). *Understanding comics: The invisible art.* New York: Harper Collins.

McCrindle, R. J., & Symons, D. *Audio space invaders.* Paper presented at 3rd International Conference on Disability, Virtual Reality and Associated Technologies, Alghero, Italy.

McMahan, A. (2003). Immersion, engagement, and presence: A new method for analyzing 3-D video games. In M. J. P. Wolf & B. Perron (Eds.), *The Video Game Theory Reader* (pp. 67—87). New York: Routledge.

Medal of Honor series. (1999-2006). [Computer Game]. Electronic Arts.

Metal gear solid 2: Sons of liberty. (2001). [Computer game]. Konami Computer Entertainment Japan Inc.

Metz, C. (1982). *The imaginary signifier: Psychoanalysis and the cinema* (C. Britton, A. Williams, B. Brewster & A. Guzzetti, Trans.) Bloomington: Indiana University Press.

Metz, C. (1985). Aural objects. In E. Weis & J. Belton (Eds.), *Film Sound: Theory and Practice* (pp. 154—161). New York: Columbia University Press.

Miyamoto, S. (1981). Donkey kong. [Computer Game]. Nintendo.

Morris, S. (2002). First person shooters -- a game apparatus. In G. King & T. Krzywinska (Eds.), *Screenplay: Cinema/Videogames/Interfaces* (pp. 81—97). London: Wallflower Press.

Murphy, D., & Pitt, I. (2001). Spatial sound enhancing virtual story telling. *Lecture Notes in Computer Science, 2197*, 20—29.

Murray, J. H. (2000). *Hamlet on the holodeck: The future of narrative in cyberspace.* Cambridge: The MIT Press.

PainBerry. (n.d.). Realism: Is it possible in any quake game? Retrieved June 2, 2006, from http://planetquake.gamespy.com/View.php?view=Editorials.Detail&id=110.

Palmer, I. J. (2002). *How realistic is the sound design in the D-Day landing sequence in Saving Private Ryan?.* Unpublished master's thesis, Bournemouth

University.
Parkes, D. N., & Thrift, N. J. (1980). *Times, spaces, and places: A chronogeographic perspective*. New York: John Wiley & Sons.
Pearsall, J. (Ed.). (2002). *Concise Oxford English dictionary* (10th ed.) Oxford: Oxford University Press.
Peep. (n.d.). Retrieved May 12, 2003, from http://peep.sourceforge.net/docs/peep_proposal.html.
PHP bibtex database manager. (2006). (Version 1.3) [PHP].
Pine, B. J., & Gilmore, J. H. (1999). *The experience economy: Work is theatre & every business a stage*. Boston: Harvard Business School Press.
Pozzi, D. (2004). PHPBibMan. (Version 0.4) [PHP].
Pudovkin, V. I. (1934). Asynchronism as a principle of sound film. Retrieved January 26, 2006, from http://lavender.fortunecity.com/hawkslane/575/asynchronism.htm.
Puterbaugh, J. (1999). Sonopoietic space. Retrieved July 7, 2006, from http://www.music.princeton.edu/~john/sonopoietic_space.htm.
Quake III Arena review. (n.d.). Retrieved April 14, 2006, from http://www.makeitsimple.com/gaming/game_reviews/quake3_arena/.
Röber, N., & Masuch, M. (2004, July 6—9). *Interacting with sound: An interaction paradigm for virtual auditory worlds*. Paper presented at Tenth Meeting of the International Conference on Auditory Display, Sydney, Australia.
Röber, N., & Masuch, M. (2005, June 16—20). *Playing audio-only games: A compendium of interacting with virtual auditory worlds*. Paper presented at Changing Views -- Worlds in Play, Toronto.
Rebelo, P. 2003. *Performing space*. Baden-Baden: Symposium on Systems Research in the Arts: Music, Environmental Design, and the Choreography of Space.
Rocket arena. (2000-2006). [Computer Game].
Roudavski, S., & Penz, F. (2003, November 4—6). *Space, agency, meaning and drama in navigable real-time virtual environments*. Paper presented at Level Up, Utrecht Universiteit.
Ryan, M.-L. (2001). Beyond myth and metaphor: The case of narrative in digital media. *Game Studies, 1*(1). Retrieved July 25, 2006, from http://www.gamestudies.org/0101/ryan/.
Sánchez, J., & Lumbreras, M. (1999, May 15). *Interactive 3d sound hyperstories for blind children*. Paper presented at Conference on Human Factors in Computing Systems, Pittsburgh, USA.
Schafer, R. M. (1994). *The soundscape: Our sonic environment and the tuning of the world*. Rochester Vt: Destiny Books.
Schott, G., & Kambouri, M. (2003). Moving between the spectral and material plane: Interactivity in social play with computer games. *Convergence, 9*(3), 41—55.
Shinn-Cunningham, B. G., Lin, I.-F., & Streeter, T. (2005, July 22—27). *Trading directional accuracy for realism in virtual auditory display*. Paper presented at 1st. International Conference on Virtual Reality.
Silicon Ice. (2005). Urban Terror. (Version 3.7) [Computer Game].
Spasim. (1974). [Computer Game].
Spielberg, S. (Director) (1998). *Saving private Ryan* [Film]. S. Spielberg, M. Gordon, G. Levinsohn & I. Bryce (Producers). USA: Dreamworks SKG.
State University Center of Creative and Performing Arts. 1968. In C. (T. Riley (Comp.)). Sony. [Vinyl Record] 7178.
Steffens, M., & Karnesky, R. (2006). Refbase. (Version 0.9) [PHP].

Stockburger, A. (2003, November 4—6). *The game environment from an auditive perspective.* Paper presented at Level Up, Utrecht Universiteit.
Taylor, L. N. ((2002). *Video games: Perspective, point-of-view, and immersion.*). Unpublished master's thesis, University of Florida.
Taylor, L. N. (2003). When seams fall apart: Video game space and the player. *Game Studies, 3*(2). Retrieved March 15, 2006, from http://gamestudies.org/0302/taylor/.
Théberge, P. (1989). The 'sound' of music: Technological rationalization and the production of popular music. *New Formations, 8*, 99—111.
Thom, R. (1999). Designing a movie for sound. Retrieved September 29, 2003, from http://www.filmsound.org/articles/designing_for_sound.htm.
Thompson, K., & Bordwell, D. (2003). *Film history: An introduction* (2nd ed.) New York: McGraw-Hill.
Truax, B. (2001). *Acoustic communication* (2nd ed.) Westport, Conn: Ablex.
Truppin, A. (1992). And then there was sound: The films of Andrei Tarkovsky. In R. Altman (Ed.), *Sound Theory Sound Practice* (pp. 235—248). New York: Routledge.
Tsingos, N., Gallo, E., & Drettakis, G. (2003). Breaking the 64 spatialized sources barrier. *Gamasutra.* Retrieved September 19, 2004, from http://www.gamasutra.com/resource_guide/20030528/tsingos_pfv.htm.
Turcan, P., & Wasson, M. (2003). *Fundamentals of audio and video programming for games.* Redmond: Microsoft Press.
Valve Software. (1998). Half-life. [Computer Game]. Sierra Studios.
Valve Software. (1999). Counter-strike. [Computer Game]. Vivendi Universal.
Valve Software. (2004). Half Life 2. [Computer Game]. Electronic Arts.
Velleman, E., van Tol, R., Huiberts, S., & Verwey, H. (2004). 3d shooting games, multimodal games, sound games and more working examples of the future of games for the blind. *Lecture Notes in Computer Science, 3118*, 257—263.
Walker, R. (1987). The effects of culture, environment, age, and musical training on choices of visual metaphors for sound. *Perception & Psychophysics, 42*(5), 491—502.
Ward, P. (2002). Videogames as remediated animation. In G. King & T. Krzywinska (Eds.), *Screenplay: Cinema/Videogames/Interfaces* (pp. 122—135). London: Wallflower Press.
Warner, T. (2003). *Pop music — Technology and creativity: Trevor Horn and the digital revolution.* Aldershot: Ashgate.
Warren, D. H., Welch, R. B., & McCarthy, T. J. (1982). The role of visual-auditory "compellingness" in the ventriloquism effect: Implications for transitivity among the spatial senses. *Perception & Psychophysics, 30*(6), 557—564.
Warren, W. H., & Verbrugge, R. R. (1984). Auditory perception of breaking and bouncing events: A case study in ecological acoustics. *Journal of Experimental Psychology: Human Perception and Performance, 10*(5), 704—712.
Wasik, B. (2006, September). Grand theft education: Literacy in the age of video games. *Harper's Magazine*, 31—39.
Watamaniuk, J. (n.d.). Neverwinter Nights: Interview with composer David John. Retrieved February 10, 2006, from http://nwn.bioware.com/players/profile_david_john.html.
Weis, E., & Belton, J. (Eds.). (1985). *Film sound: Theory and practice.* New York: Columbia University Press.
Wenzel, E. M. (1992). Localization in virtual acoustic displays. *Presence, 1*(1), 80—

107.
Westerkamp, H. (2000). Editorial. *Soundscape, 1*(1), 3—4.
Whalen, Z. (2004). Play along -- an approach to videogame music. *Game Studies, 4*(1). Retrieved February 20, 2005, from http://www.gamestudies.org/0401/whalen/.
Witkin, H. A., Wapner, S., & Leventhal, T. (1952). Sound localization with conflicting visual and auditory cues. *Journal of Experimental Psychology, 43*, 58—67.
Wolf, M. J. P. (2001a). Space in the video game. In M. J. P. Wolf (Ed.), *The Medium of the Video Game* (pp. 51—75). Austin: University of Texas Press.
Wolf, M. J. P. (2001b). Time in the video game. In M. J. P. Wolf (Ed.), *The Medium of the Video Game* (pp. 78—91). Austin: University of Texas Press.
Wurtzler, S. (1992). "She sang live, but the microphone was turned off": The live, the recorded and the subject of representation. In R. Altman (Ed.), *Sound Theory Sound Practice* (pp. 87—103). New York: Routledge.
Zhang, J. (n.d.). Categorization of affordances. Retrieved March 20, 2006, from http://acad88.sahs.uth.tmc.edu/courses/hi6301/affordance.html.
Zizza, K. (2000). Your audio design document: Important items to consider in audio design, production, and support. *Gamasutra*. Retrieved September 18, 2003, from http://www.gamasutra.com/features/20000726/zizza_01.htm.